Ernst Schering Research Foundation Workshop 36
The Human Genome

Springer

Berlin
Heidelberg
New York
Barcelona
Hong Kong
London
Milan
Paris
Tokyo

Ernst Schering Research Foundation Workshop 36

The Human Genome

Biology and Medicine

A. Rosenthal, L. Vakalopoulou
Editors

With 32 Figures and 12 Tables

 Springer

Series Editors: G. Stock and M. Lessl

Learning Resources
Centre
1 2 3 2 3 0 7 1

ISSN 0947-6075
ISBN 3-540-42316-8   Springer-Verlag Berlin Heidelberg New York

Die Deutsche Bibliothek - CIP-Einheitsaufnahme
The human genome : biology and medicine / A. Rosenthal and L. Vakalopoulou ed.. - Berlin ;
Heidelberg ; New York ; Barcelona ; Hongkong ; London ; Milan ; Paris ; Singapore ; Tokyo : Sprin-
ger, 2002
  (Ernst Schering Research Foundation Workshop ; 36)
  ISBN 3-540-42316-8

Springer-Verlag Berlin Heidelberg New York
a member of BertelsmannSpringer Science+Business Media GmbH

http://www.springer.de

© Springer-Verlag Berlin Heidelberg 2002
Printed in Germany

Typesetting: Data conversion by Springer-Verlag
Cover design: design & production, Heidelberg
Printing: Druckhaus Beltz, Hemsbach
Binding: J. Schäffer GmbH & Co. KG, Grünstadt
SPIN: 10844498   21/3130/AG 5 4 3 2 1 0 Printed on acid-free paper

Preface

In recent months much has been said about the Human Genome Project (HGP). Some people have compared the achievement with the landing of human beings on the moon, others with the invention of the printing press. Certainly the human genome sequence in the context of genetic programs from other organisms like the fruit fly *D. melanogaster* and the round worm *C. elegans* is a scientific achievement which is unparalleled. The consequence for science, especially for biology and medicine, but also for our society in general, are manifold, profound, and at the moment hardly understood. Nothing will change our way of life in the future more than the effects of this revelation.

Sequencing just the working draft of the entire human genome was technically an extremely challenging job, involving thousands of people working for over five years in six different countries. The competition between the international academic consortium and Celera Genomics helped deliver the genome much earlier than expected. In December 1999, the first human chromosome – chromosome 22 – was completed by an international consortium led by the British Sanger Centre. This effort was followed in May 2000 by chromosome 21 which was sequenced by a German/Japanese consortium. Both chromosomes were published in "Nature" and will very likely serve as the golden standard for all other chromosomes with respect to the depth of the physical maps and the quality of the final sequence. For example, chromosome 21 retains only three physical gaps comprising only 100 kb or just 0.3% of the long arm, while 3% of chromosome 22 is still missing. The next human chromosomes to be finished in 2002 will

The participants of the workshop

very likely be chromosomes 14 and 20. In June 2000 the working draft
of the entire human genome was announced followed by it's publication as a whole in "Nature" on February 12, 2001. Celera Genomics
published their version of the genome one day later in "Science". What
counts at the end of the day, however, is not the competition, but the
sequence of the human genome itself in the highest possible quality or-
ganised in single chromosomes and accessible for each individual or
public or private institution with no fees or conditions attached. We,
the organizers of this workshop, are very proud that leading repre-
sentatives of the international teams involved in the sequencing of
chromosomes 22, 21 and 14, among them Ian Dunham and Jean Weissenbach reviewed some of the primary data sets.

In the last few months billions of human bases have been flooding
databases. Soon we will be overwhelmed by even more sequences
originating from important mammalian and vertebrate model organisms like mouse and rat, zebrafish and primates. Who is prepared for
this excess of genetic information? Except for a very few large bioin-
formatics centres most research groups are lost in this ocean. Also,
many pharmaceutical companies cannot deal with the information deluge. At the time of the workshop only 20% of the genome existed in

high quality while a year later in October 2001 about 50% of the genome is finished. Half of the genome, however, is still in working draft quality. This draft is often highly fragmented and not necessarily useful to predict novel genes, e. g. genes for which no mRNA is available, or genes lacking any detectable similarity to protein or protein motifs from other organisms.

The annotation of the chromosome 21 and 22 sequence suggested that the human genome very likely encodes only approximately 30.000 to 40.000 genes, much fewer than originally anticipated. This hypothesis, first announced in May 2000 at the Cold Spring Harbor Symposium on Genome Sequencing and Biology was supported by other data sets from the puffer fish sequencing projects as well as by analysis of the EST data. Finding most human genes within the draft sequence though is still a complicated and demanding problem. Currently, we have a good understanding of about 12.000 human genes for which full length or near full length mRNA's are available. Another approx. 10.000 gene models can be predicted from the draft sequence due to their high similarity to proteins from a range of model organisms. These *in-silico* gene models are incomplete and need to be confirmed by experimental methods. However, 10.000 to 15.000 human genes cannot reliably be predicted from the current working draft. Those people who can make sense of the working draft by comparing it with other data sets from mouse and rat will be the mid and long term winners in the hunt for novel proteins. Hugues Roest Crollius convincingly showed how the random sequence from the evolutionary distant puffer fish Tetraodon can be used as a tool to verify exons predicted *in-silico* in man and to reveal human gene number, distribution and rearrangements.

The HGP involved much more than simply sequencing the 3 billion or so nucleotides which make up the genetic program of a human cell. The really tough question is however: how do we go from sequence to function and then to disease, or from sequence to disease and then to function? This question needs to be addressed for the approximately 30.000-40.000 different human genes, and perhaps 150.000 to 200.000 different transcripts. Due to differential splicing each transcript may be translated into a different protein. In addition, each protein may be further post-translationally modified by sugar or phosphate moieties making the picture even more complex. Thus, while there may be half a million different proteins, each human cell type may only produce different protein subsets. Furthermore, these genes and their products (proteins or RNAs) operate in thousands of interconnected and cascading networks.

Traditional genetics addresses the function of an isolated gene by linking it to a mutant phenotype. In man, positional cloning combined with transgenesis have been powerful strategies to study some aspects of gene function. Positional cloning has allowed the identification of more than 1.000 disease genes over the last 15 years. It often took several years untill a particular disease gene had been successfully identified. Following their identification many mutations have been described for each of those genes, and human geneticists tried to correlate the mutations with distinct phenotypes. In addition, single genes can be knocked out or over expressed in almost any model organism or cell line giving clues of the putative function of a particular disease gene. Some of these key questions have been addressed during the workshop and are now summarised in the book. Yossi Shiloh, who in 1995 reported the positional cloning of the gene involved in Ataxia Telangiectasia, reviewed what we presently know about the function of the ATM gene. He also addressed the ways to go from disease to sequence, and then to function and back to disease. Karen Avraham gave an overview about the molecular genetics of deafness – a complex trait. Sue Kenwrick reported on the positional cloning of the NF-kappa B gene involved in Intercontinentia Pigmenti. And Pat Nolen showed the power of ENU mutagenesis in creating new mouse models for human diseases.

Understanding the precise role and function of the majority of human genes in complex diseases, like sporadic cancers for instance may take a long time. However, predisposition to disease and side reaction to drug treatment are determined by a large number of unique genetic variations contained in our genome. There are 3 million polymorphic bases in our genome. However, only approximately 150.000 to 300.000 of these single nucleotide polymorphisms (SNPs) are located within genes and even fewer may alter amino acid positions or the expression of a particular gene. Using high density SNP arrays the variations between individuals and populations can be obtained at the germline level and then linked with clinical data of patients with complex diseases. Thus, patient profiling studies have the potential to relate the functional response of an individual or a specific population to their genotype. While the medicine of the future will include new therapies and diagnostic tools, it will also be predictive and tailored to the individual genotype. Populations can be stratified to identify key pathways for the design and application of novel therapies. Depending on the complex genotype of a patient, conventional therapies can be chosen such that side reactions are minimized and longevity increased. Thus, new strategies in diagnostic, prevention and therapy will lead to a radical change of medicine.

It is our very great honour and pleasure to especially thank Jean Weissenbach – the key speaker of this symposium for his opening lecture: "The human genome project: past, present, future". Finally we would like to thank the Ernst Schering Research Foundation, especially Dr. Monika Lessl, but also Mrs. Szivos, Mrs. Wanke, and Mrs. Elsinghorst for having helped to organise such a wonderful meeting and supporting the preparation of the proceedings.

André Rosenthal
Lilian Vakalopoulou

Contents

List of Editors and Contributors

Editors

Rosenthal, A.
MetaGen Gesellschaft für Genomforschung mbH, Oudenarder Straße 16,
13347 Berlin, Germany

Vakalopoulou, L.
MetaGen Gesellschaft für Genomforschung mbH, Oudenarder Straße 16,
13347 Berlin, Germany

Contributors

Avraham, K.B.
Department of Human Genetics and Molecular Medicine;
Sackler School of Medicine, Tel Aviv University, Ramat Aviv,
Tel Aviv 69978, Israel

Bernot, A.
Genoscope and CNRS UMR8030, 2 rue Gaston Crémieux,
91057 Evry Cedex, France

Bouneau, L.
Genoscope and CNRS UMR8030, 2 rue Gaston Crémieux,
91057 Evry Cedex, France

Burge, C.
Massachusetts Institute of Technology, Department of Biology, 68-222,
Cambridge, MA 02139-4307, USA

Dasilva, C.
Genoscope and CNRS UMR8030, 2 rue Gaston Crémieux,
91057 Evry Cedex, France

Dunham, I.
Sanger Centre, Wellcome Trust Genome Campus, Hinxton,
Cambridge CB11 4TG, UK

Hardisty, R.
MRC Mammalian Genetics Unit and Mous Genome Centre, Harwell, Didcot,
Oxfordshire OX11 0RD, UK

Hough, T.
MRC Mammalian Genetics Unit and Mous Genome Centre, Harwell, Didcot,
Oxfordshire OX11 0RD, UK

Hunter, A.J.
Department of Neuroscience, SmithKline Beecham Pharmaceuticals,
New Frontiers Science Park, Harlow, UK

Jaillon, O.
Genoscope and CNRS UMR8030, 2 rue Gaston Crémieux,
91057 Evry Cedex, France

Kenwrick, S.
Cambridge Institute for Medical Research, Addenenbrooke's Hospital,
Hills Road, Cambridge CB2 2XY, UK

Nolan, P.M.
Medical Research Council, Harwell, Didcot, Oxfordshire OX11 0RD, UK

Pelletier, E.
Genoscope and CNRS UMR8030, 2 rue Gaston Crémieux,
91057 Evry Cedex, France

Quetier, F.
Genoscope and CNRS UMR8030, 2 rue Gaston Crémieux,
91057 Evry Cedex, France

Roest Crollius, H.R.
Genoscope and CNRS UMR8030, 2 rue Gaston Crémieux,
91057 Evry Cedex, France

Saurin, W.
Genoscope and CNRS UMR8030, 2 rue Gaston Crémieux,
91057 Evry Cedex, France

Shiloh, Y.
Department of Human Genetics and Molecular Medicine,
Sackler School of Medicine, Tel Aviv University, Ramat Aviv,
Tel Aviv 69978, Israel

Thaung, C.
MRC Mammalian Genetics Unit and Mouse Genome Centre, Harwell,
Didcot, Oxfordshire OX11 0RD, UK

Weissenbach, J.
Genoscope and CNRS UMR8030, 2 rue Gaston Crémieux,
91057 Evry Cedex, France

Yeh, R-F.
Massachusetts Institute of Technology, Department of Biology, 68-222,
Cambridge, MA 02139-4307, USA

1 Human Genome Project: Past, Present, Future

J. Weissenbach

1 A Short History of the Human Genome Project

The Human Genome Project was launched in the United States in September 1990. Several other human genome programmes were started in the early 1990s in other countries. The first necessity was to establish maps covering the entire genome, both as a tool for medical purposes to facilitate identification of genes involved in genetic diseases and to set up the infrastructure for the sequencing phase.

1.1 Mapping Phase (1990–1998)

The genetic map that has been established over the past decade currently (Dib et al. 1996) comprises a total of about 16,000 genetic markers of which about 12,000 are highly informative second generation markers of the microsatellite type. At present, the density of markers on the genetic map is sufficient to localize easily a gene for a monogenetic disease with a precision of 1–2 million base pairs, and to define intervals which contain susceptibility genes for multifactorial disorders. A third generation map is in progress. It is based on single nucleotide polymorphisms (SNPs) that are much less informative than microsatellites from the second generation map, but can be genotyped using DNA chip technology or other high throughput procedures currently under development. More than 1,500,000 SNPs have been developed by a consortium which includes public genome centres and about ten of the major pharmaceutical companies (http://snp.cshl.org/). These markers have also been placed on the sequence of the working draft.

A physical map, based on sets of overlapping yeast artificial chromosomes (YACs) ordered on the basis of their content of sequence-tagged sites (STSs), has also been established (Hudson et al. 1995). However, this physical map could not serve as a support for sequencing because of the numerous rearrangements that YAC clones have undergone. The map based on YACs is now replaced by a new physical map based on bacterial artificial chromosomes (BACs), which are much less subject to rearrangements. These BACs have been assembled by the Genome Centre of Washington University at St. Louis in sets of overlapping segments using restriction digest fingerprinting (International Human Genome Mapping Consortium 2001). BAC clones representing a minimal tiling path of overlapping DNA segments and covering about 97% of the euchromatic part of the genome have further served for the establishment of the working draft sequence (see below). Conversely, the sequence has served to validate and occasionally correct the fingerprint based BAC contig map. Additional BAC clones that extend the sequenced contigs and fill the remaining gaps are currently being added to the existing map and sequence.

Several collections of whole genome radiation-induced hybrids have been developed during the mid-1990s. They were particularly useful for integrating genes with existing maps. Using these collections, a network of American and European laboratories has mapped a set of more than 30,000 expressed sequence tags (ESTs) derived from cDNA libraries prepared for large-scale sequencing programmes (Deloukas et al. 1998).

1.2 Sequencing Phase (1996–2003)

The sequencing phase was launched at a pilot level in 1995–1997. It has gone through several steps of acceleration which culminated with the recent completion of the intermediary step known as the working draft. This sequencing programme is carried out by an international consortium of publicly funded sequencing centres. Most of the funding is managed by the National Institutes of Health, the Wellcome Trust and the Department of Energy and to a lesser extent by similar governmental funding in China, Germany, Japan and France. The strategy of the publicly funded programme has been reformulated in 1999. In the first step of the sequencing phase, a sequence known as the "working draft" was recently established with low-depth genome coverage (4–5 genome equivalents) by shotgun sequencing of large overlapping fragments (cloned in BACs) that are ordered and cover the ensemble of the genome (see Sect. 1.1). The sequencing has been divided up among various centres at the international level, and the resulting data were made publicly available as they were established in sequencing centres. Twenty centres participated in this working draft effort (International Human Genome Sequencing Consortium 2001).

The raw data from the different BAC clones have been assembled into an integrated draft sequence of the human genome. A computer-based process ordered and oriented the sequence contigs using various sources of mapping information. The sequence contigs from different BACs could be joined together when overlapping or linked together to create scaffolds.

As of march 2001 the amount of non-redundant finished sequence that is publicly available represents about 850 Mb and the part in draft another 1,850 Mb. There remain about 140,000 sequencing gaps, 4,000 gaps between sequenced clone contigs and 1,000 gaps between mapped clone contigs.

This first step will be followed by an additional shotgun sequencing of the same fragments to reach a coverage of ten genome equivalents. Although this second step will permit ordering and orientation of the vast majority of sequence contigs assembled, this does not represent a completed sequence of the human genome. This latter will require an additional finishing step which will lead to the complete sequence of the euchromatic part of the human genome with practically no gaps by the year 2003. Finished sequences of two entire chromosomes have been published in the past 2 years, and in some way they set the standard of the finished human sequence (Dunham et al. 1999; Hattori et al. 2000).

On the other hand, a private company (Celera Genomics) has been sequencing the human genome using a completely random sequencing strategy which does not depend on the prior establishment of a map of ordered clones. It is, however, difficult to evaluate the success of this strategy since Celera has now merged the primary data set produced for the "working draft" and its own shotgun reads which represent another 5× coverage (Venter et al. 2001). Again this combination does not constitute a completed sequence of the human genome. It provides only marginal extension of coverage, and no fewer sequence gaps. Some assembly errors from the public project could be identified. Unfortunately, Celera provides no information on the assembly uniquely based on the data set they produced. It is probable that the amount of sequence produced by Celera was not sufficient per se to achieve a valuable assembly.

2 Using the Human Genome Sequence

At the present time, the deluge of sequence data has mainly served to reveal our ignorance of biological reality, and the analysis itself of this data is far from complete due to a growing series of complications:

– We still encounter difficulties in precisely identifying genes in the DNA sequence, especially in higher eukaryotes.
– We are rarely capable of attributing function to a newly identified gene with certainty.
– We are practically unable to predict the phenotype which would result from an alteration of an even known function in a higher eukaryote.

2.1 Defining the Genes

Since it is a prerequisite for more functional studies, the first priority was to define genes on the genomic sequence. The difficulties of this exercise have been underestimated and the present estimates of number of genes are still not definitive. However, the gene counts that were recently published (International Human Genome Sequencing Consortium 2001; Venter et al. 2001), are in the same range as estimates that were made about a year before (Ewing and Green 2000; Roest Crollius et al. 2000).

One cause of that difficulty resides in the fact that the mean size of 20 kb of assembled DNA fragments in the unfinished fraction of the sequence is smaller than the mean size of a human gene (27 kb). Furthermore, neighbouring fragments are only rarely ordered and oriented in the present stage of the working draft. There are, in addition, some difficulties in finding genes on the finished sequence as well.

Two main types of methods are used to define genes. The first is based on searching sequence similarities or identities in other data sets and the second relies on the use of programmes that predict characteristic gene features. Similarity searches are based on the use of powerful sequence comparison tools like basic local alignment search tool (BLAST) (Altschul et al. 1990). Although BLAST searches are very efficient, they suffer a major limitation due to the quality of datasets that are used for comparisons. Typically, a mammalian genome like the human genome would be most suited for comparisons with other mammalian genomes that are sufficiently evolutionarily distant, like the mouse. The rapidly growing amount of genomic sequence data from mouse should, however, soon provide a critical set for gene annotation. Although other animal genomes, like *Drosophila* and *Caenorhabditis* are very helpful, there are genes that have only appeared recently in evolution and others that drifted apart much faster and that are no longer sufficiently similar between vertebrates and non-vertebrates. Comparisons to cDNAs and ESTs, from either the cognate or related species are also essential. However, the EST and cDNA sequence collections that are publicly available to date remain very incomplete, and conversely contain also parts of dubious transcripts. These collections have been tentatively ordered, using clustering procedures based on sequence identities (Schuler 1997). Such procedures usually end up with more than one cluster per gene, suggesting that alternative splicing occurs frequently and is possibly one of the major mechanisms allowing diversity and flexibility.

The exon and gene prediction programs perform rather satisfactorily and are capable of detecting most coding regions (Claverie 1997). Their major drawbacks reside in a lack of accuracy for gene delimitation and in a frequent tendency to overpredict exons on the genomic sequence.

2.2 Gene Function

The classical procedure in genetics has usually consisted in deriving a genotype from a phenotype. The availability of complete genome sequences provides an invaluable tool to facilitate research in genetics and especially to find genes responsible for defined phenotypes. In many of the most recent searches for genes responsible for diseases, such genes were found with the help of the genomic sequence. This first outcome is far from being exhausted yet, but it can be expected that with the availability of a first draft of the human genome sequence most of the human Mendelian disease genes will be found in the forthcoming years, provided they can be successfully mapped by linkage analysis.

It was initially hoped that the concepts and tools used to identify the cause of Mendelian diseases could also be applied to the more common diseases. However, the gene mutations involved in Mendelian diseases are directly the cause of the disease, whereas gene alterations involved in multifactorial diseases are only predisposing factors which exert their effect in only a fraction of the carriers. Because of these differences, the search for common disease susceptibility genes is much more difficult and based on different strategies and tools.

Although we all have identical genomes there are some small differences (single nucleotide polymorphisms or SNPs) distributed at random at a mean frequency of one change per 1,000 bases between two homologous chromosomes. Rarely these alterations (possibly even a single event which occurred many generations ago) may modify a gene to become predisposing for a common disease. Such alterations have occurred in a given context, surrounded by a set of SNPs that is characteristic of that context. Although after many generations the process of recombination will progressively reduce the size of the original segment, the SNPs closest to the predisposing alteration will remain in their original form. We will thus be able to trace this original chromosome segment in disease-affected individuals by looking at the SNPs that show a statistically significant association with the disease versus control subjects.

Because the original segments in which the predisposing alteration occurred will become very reduced in size after many generations, it will be necessary to scan the genome at a very high density to be able to detect the SNPs associated with the disease. In addition, to reach statistically significant associations we will have to screen very large numbers of both affected and control individuals (cohorts).

It is for these reasons that DNA collections from large cohorts of affected and control individuals are being established and that several very large-scale programs to develop large numbers of SNPs are in progress. More than 1 million SNPs are available in public databases, and even larger numbers can be found in private databases, and these collections are still increasing. Large-scale screenings of these large cohorts necessitates the development of methods that will enable investigators to genotype and analyse massive numbers of SNPs on large populations.

Despite several major difficulties (genetic heterogeneity within a gene – due to multiple founder effects – or between different loci, influence of other factors etc.), the finding of predisposing genes will accumulate, although at an unpredictable pace, which may turn out to be rather slow. The tools to find genes which predispose to common diseases are now at hand but will continually be improved with the foreseeable completion of the human genome sequence and increase of SNP collections.

These resource will be completed by other approaches based on functional genomics. Gene or protein function will still be studied by all the classical approaches that have been developed in the past. Given the fact that these studies are usually carried out on single entities, major bottlenecks have now appeared in many traditional areas of biology. The availability of complete genome sequences also enables investigators to perform experiments taking simultaneously into account the global gene set of an organism, tissue or cell. Such studies should provide a wealth of information on gene networks, protein–protein interactions and so forth (Schena et al. 1995; Eisen et al. 1998; Enright et al. 1999; Marcotte et al. 1999a,b; Ross-Macdonald et al. 1999; Uetz et al. 2000).

Although gene function can sometimes be strongly suggested by sequence similarity searches (Marcotte et al. 1999a,b; Pellegrini et al. 1999), experimental approaches remain the only way to validate such predictions and to partially overcome limitations of the in silicio approaches. In addition, genes without any associated phenotypes are still numerous. We have a long way to go before we will be able to deduce a phenotype from a sequence and before we understand the basis of phenomena like epistasis, pleiotropy, dominance, etc. The functional

8 J. Weissenbach

genomics programs in progress or to be implemented are certainly
essential, but their impact on this problem remains uncertain.

References

Altschul SF, Gish W, Miller W, Myers EW, Lipman DJ (1990) Basic local
alignment search tool. J Mol Biol 215:403–410

Claverie JM (1997) Computational methods for the identification of genes in
vertebrate genomic sequences. Hum Mol Genet 6:1735–1744

Deloukas P, Schuler GD, Gyapay G, Beasley EM, Soderlund C et al (1998) A
physical map of 30,000 human genes. Science 282:744–746

Dib C, Fauré S, Fizames C, Samson D, Drouot N et al (1996) A comprehensive genetic map of the human genome based on 5,264 microsatellites. Nature 380:152–154

Dunham I, Shimizu N, Roe BA, Chissoe S, Hunt R et al (1999) The DNA sequence of human chromosome 22. Nature 402:489–495

Eisen MB, Spellman PT, Brown PO, Botstein D (1998) Cluster analysis and
display of genome-wide expression patterns. Proc Natl Acad Sci USA
95:14863–14868

Enright AJ, Iliopoulos I, Kyrpides NC, Ouzounis CA (1999) Protein interaction maps for complete genomes based on gene fusion events. Nature
402:86–90

Ewing B, Green P (2000) Analysis of expressed sequence tags indicates
35,000 human genes. Nat Genet 25:232–234

Hattori M, Fujiyama A, Taylor D, Watanabe H, Yada T et al (2000) The DNA
sequence of human chromosome 21. Nature 405:311–319

Hudson TJ, Stein LD, Gerety SS, Ma JL, Castle AB et al (1995) An STS-based
map of the human genome. Science 270:1945–1954

International Human Genome Sequencing Consortium (2001) Initial sequencing and analysis of the human genome. Nature 409:860–921

International Human Genome Mapping Consortium (2001) A physical map of
the human genome. Nature 409:934–941

Marcotte EM, Pellegrini M, Ng HL, Rice DW, Yeates TO, Eisenberg D (1999a)
Detecting protein function and protein-protein interactions from genome sequences. Science 285:751–753

Marcotte EM, Pellegrini M, Thomson MJ, Yeates TO, Eisenberg D (1999b) A
combined algorithm for genome-wide prediction of protein function. Nature 402:83–86

Pellegrini M, Marcotte EM, Thomson MJ, Eisenberg D, Yeates TO (1999) As-
signing protein functions by comparative genome analysis: protein phylo-
genic profiles. Proc Natl Acad Sci USA 96:4285–4288

Roest Crollius H, Jaillon O, Bernot A, Da Silva C, Bouneau L et al (2000) Estimate of gene number provided by genome-wide analysis using *Tetraodon nigroviridis* DNA sequence. Nat Genet 25:235–238

Ross-Macdonald P, Coelho PS, Roemer T, Agarwal S, Kumar A et al (1999) Large-scale analysis of the yeast genome by transposon tagging and gene disruption. Nature 402:413–418

Schena M, Shalon D, Davis RW, Brown PO (1995) Quantitative monitoring of gene expression patterns with a complementary DNA microarray. Science 270:467–470

Schuler GD (1997) Pieces of the puzzle: expressed sequence tags and the catalog of human genes. J Mol Med 75:694–698

Uetz P, Giot L, Cagney G, Mansfield TA, Judson RS et al (2000) A comprehensive analysis of protein-protein interactions in *Saccharomyces cerevisiae*. Nature 403:623–627

Venter C, Adams M, Myers E, Li P, Mural R et al (2001) The sequence of the human genome. Science 291:1304–1351

2 Genome-wide Comparisons Between Human and Tetraodon

H.R. Roest Crollius, O. Jaillon, A. Bernot, E. Pelletier, C. Dasilva,
L. Bouneau, C. Burge, R-F. Yeh, F. Quetier, W. Saurin,
J. Weissenbach

1 Introduction

The sequence of the human genome (Lander et al. 2001) provides the foundation for new approaches to study the organization and function of all human genes. An obvious prerequisite for such studies is the identification of the genes in the genome, which remains a challenging task. Initially to complement other approaches, Exofish (exon-finding by sequence homology) was developed to identify human genes based on their conservation in the pufferfish *Tetraodon nigroviridis* (Roest Crollius et al. 2000a). Here we present an annotation of the sequence of the human genome using Exofish, and compare the position of evolutionarily conserved regions (ecores) with gene annotations produced by more conventional strategies (Lander et al. 2001). The human genome sequence also provides the first opportunity to study a vertebrate genome in comparison with other genomes at the molecular level. Such studies are likely to reveal fundamental features of both the evolutionary history of the vertebrate genome and of its modern structure. We have performed the first such study by comparing the sequence of *Tetraodon* to the sequence of the human genome, as well as to the complete genome of *Drosophila melanogaster* and *Caenorhabditis elegans*.

Tetraodon is the vertebrate with the largest fraction of genome sequence in the public domain after human and mouse. The present study is based on 0.4 genome equivalents, or 175,000 sequence reads (150 Mb of DNA) generated from the ends of bacterial artificial chromosome (BAC) and plasmid clones which represent a theoretical coverage of 33% (Roest Crollius et al. 2000a). The 380 megabases (Mb) *Tetraodon* genome is composed of 21 chromosome pairs (Grutzner et al. 1999; Fischer et al. 2000) and is the smallest known vertebrate genome (Hinegardner 1968; Lamatsch et al. 2000). It is 20–30 million years distant from the marine pufferfish *Fugu rubripes* (Crnogorac-Jurcevic et al. 1997), and approximately 100 million years distant from the zebrafish *Danio rerio*. A detailed analysis of approximately 50 Mb of *Tetraodon* genomic DNA showed that the genome contains less than 10% of repetitive DNA (Roest Crollius et al. 2000b), giving rise to smaller intergenic and intronic regions compared to vertebrates with larger genomes. We used the *Tetraodon* sequence to perform comparative Exofish analyses between *Tetraodon* and *Homo sapiens*, *D. melanogaster* and *C. elegans*. According to this analysis, both vertebrate genomes contain approximately 30,000 genes. Results indicate that more than a third of human coding DNA is not yet recognized by standard gene annotation methods, whereas gene annotation in the fly and worm genomes appear much more complete. A genome-wide distribution of human ecores that share homologies to each other provides a glimpse of the evolutionary dynamics that have shaped the modern human genome.

2 Results

2.1 Exofish Analysis of the GoldenPath

We have analysed the entire GoldenPath sequence (version September 2000) identifying the position of 76,866 ecores. Their distribution in the human genome is very variable, with chromosomes 17 and 19 being particularly dense and chromosomes 4, 13 and 21 being ecore-poor relative to their sizes (Table 1). The 76,866 ecores represent coding sequences and are proportional to the number of genes in the genome. In order to derive the number of human genes from the number of ecores, we determined the average number of ecores per gene in 10,026 complete and non-redundant human genes catalogued in Refseq (Maglott et al. 2000). The National Center for Biotechnology Information's reference sequence project, RefSeq, is likely to be a good representation of all human genes because of the large size of the sample, the diverse origins of the sequences in terms of tissue and time of expression but also in terms of the diverse identification techniques used. In addition, each RefSeq sequence is derived from a confirmed mRNA molecule and, therefore, represents an expressed gene. Exofish detects on average 2.7 ecores in RefSeq genes (confidence interval: [2.62;2.78], $\alpha=0.05$) and 65% of the genes are identified by at least one ecore. This provides a measure of its sensitivity in detecting complete human protein coding genes based on the current 0.4 genome equivalents of the *Tetraodon* genome.

We used the average number of ecores per gene determined using RefSeq to estimate the number of genes on each human chromosome from their ecore content (Table 1). We then summed the genes predicted on each chromosome to give a total amount of 30,380 genes. This number is in agreement with our earlier estimate based on 42% of the human working draft that predicted 28,000–34,000 genes (Roest Crollius et al. 2000a). The ~42% of genome sequenced at the time of this previous estimate showed a much higher ecore density (33.1 ecores/Mb) than the ~45% of human genome sequenced in the year 2000 only (24.9 ecores/Mb). This may due to the early targeting of expressed sequence tag (EST)-containing BAC clones for sequencing (so-called "seed clones") which left a higher proportion of gene-poor clones for the later stages of the sequencing project. Our estimate of the number of human

Table 1. Distribution of ecores and estimated number of genes on human chromosomes in the September 2000 version of the GoldenPath

Human Chromosome	Estimated size (Mb)	Sequenced DNA in GoldenPath (Mb)	Number of ecores on sequenced DNA	Number of ecores on estimated size	Ecore density versus whole genome	Estimated number of genes
1	263	207.647	7,454	9,441	1.26	3,112 (±92)
2	255	216.985	5,693	6,690	0.92	2,205 (±65)
3	214	182.936	4,734	5,537	0.91	1,825 (±54)
4	203	161.954	2,754	3,451	0.60	1,137 (±34)
5	194	166.872	4,136	4,808	0.87	1,584 (±47)
6	183	160.526	3,817	4,351	0.83	1,434 (±43)
7	171	143.360	3,574	4,263	0.88	1,405 (±42)
8	155	125.853	2,520	3,103	0.70	1,023 (±30)
9	145	103.130	2,658	3,737	0.91	1,231 (±37)
10	144	123.491	3,341	3,895	0.95	1,284 (±38)
11	144	129.732	4,226	4,690	1.14	1,546 (±46)
12	143	123.084	4,495	5,222	1.28	1,721 (±51)
13	98	90.482	1,492	1,615	0.58	532 (±16)
14	93	88.307	2,358	2,483	0.94	818 (±24)
15	89	72.283	2,799	3,446	1.36	1,136 (±34)
16	98	72.368	2,646	3,583	1.28	1,181 (±35)
17	92	71.709	4,467	5,730	2.19	1,889 (±56)
18	85	73.825	1,496	1,722	0.71	567 (±17)
19	67	53.275	4,221	5,308	2.78	1,749 (±52)
20	72	64.166	2,024	2,271	1.11	748 (±22)
21	34	33.824	582	585	0.60	192 (±6)
22	34.491	33.786	1,571	1,603	1.63	528 (±16)
X	164	130.658	2,803	3,518	0.75	1,159 (±34)
Y	35	21.246	209	344	0.34	113 (±3)
NA	0	7.480	327	327	1.53	107 (±3)
UL	0	40.453	469	469	0.41	154 (±5)
Total	3,175.491	2,699.432	76,866	92,192	1.00	30,380 (±902)

For each chromosome, the total number of genes was calculated as follows: the number of ecores on the sequenced DNA was extrapolated to the full chromosome based on its estimated size, minored by 11% to eliminate the fraction of ecores that is estimated to fall in pseudogenes according to Roest Crollius et al. (2000a), and divided by 2.7, the average number of ecores per human gene

NA, genomic DNA from clones that could not be organized in contigs; UL, genomic DNA sequences from clones placed in contigs that could not be assigned to a chromosome (see Lander et al. 2001 for details)

genes is not affected by the overall lower ecore density in the human genome, because independently the average number of ecores per gene has now been refined to 2.7 instead of the 3.18 estimated previously on a smaller set of genes (Roest Crollius et al. 2000a).

Chromosomes 22 and 21 are the only two chromosomes that have been completely sequenced to date and that have been annotated, with 545 and 226 genes respectively. In comparison, Exofish predicts that there should be 528 genes on chromosome 22 and 192 genes on chromosome 21, or respectively 5% and 15% fewer genes than have been annotated. We believe that the lower Exofish estimates reflect the nature of the initial annotations, where a number of genes were fragmented as one 5' annotation (based on homologies) and one 3' annotation (based on EST matches), and hence, they will eventually merge and lower the final number of annotated genes. The sex-specific part of chromosome Y is predicted to contain 113 genes. However, chromosome Y is notorious for its high level of pseudogenes, which strongly deviates from the ~20% observed on chromosome 22 (and chromosome 21) that we used to estimate this parameter (Roest Crollius et al. 2000a). It is, therefore, likely that the true number of genes on chromosome Y is much lower, since many of its 209 ecores will fall in pseudogenes.

2.2 Gene Annotations in Sequenced Metazoan Genomes

Exofish can be used to determine the depth at which genes have been annotated in a genome. *D. melanogaster* and *C. elegans* are two other metazoans for which most of the genome sequence is available together with in-depth annotations (Adams et al. 2000; Lander et al. 2001). Their respective evolutionary distance with *Tetraodon* is greater than that between *Tetraodon* and *H. sapiens*, which makes it possible to use Exofish on these genomes without further calibration. Indeed, previous calibrations on human chromosome 22 (2) show that less than 1% of ecores represent conservation of non-coding DNA between *H. sapiens* and *Tetraodon* (false positives). Because of the greater evolutionary distance, it is, therefore, even less likely that any will be detected between *Tetraodon* and either *D. melanogaster* or *C. elegans*. For each organism, we first determined the number of ecores in the genome, and then the proportion of these previously annotated (Table 2). The fraction

Table 2. Summary of Exofish analysis of the *H. sapiens*, *D. melanogaster* and *C. elegans* sequenced genomes and respective gene annotations. RefSeq ecores were directly identified by comparison to *Tetraodon* DNA, while the positions of GenomeScan and Ensembl annotations were directly compared to the positions of ecores, both in the July 2000 version of the GoldenPath

	H. sapiens		D. melanogaster		C. elegans
Genomic DNA compared to *Tetraodon* by Exofish (Mb)	2699.432		116.1		99.3
Number of ecores detected by Exofish in genomic DNA ("genomic ecores")	76,866		9,227		6,581
Genome annotations or gene catalogues	Genome Scan	Ensembl	RefSeq	Adams et al. (2000)	Wormpep 20
Number of annotations in dataset	42,543	35,500	10,026	13,607	19,101
Number of ecores in annotations	50,096	35,584	27,086	8,913	6,536
Percentage of annotations identified by Exofish	42.8%	36.9%	65.1%	24.7%	14.8%
Percentage of "genomic ecores" in annotations	65.17%	46.3%	35.2%	96.6%	99.3%
Mean number of ecores per annotation	1.18	1.0	2.70	0.65	0.34

of ecores present in the genomic DNA but not in the annotations is expected to be an indication of the amount of coding DNA that remains to be annotated. In *D. melanogaster* and in *C. elegans*, respectively, 96.6% and 99.3% of the ecores identified in the genome are present in the annotated genes. According to this analysis, it appears that in both cases the annotations capture nearly all the coding content detected by Exofish, and probably all the coding content of these genomes. In retrospect, this also confirms the very low background of false-positive predictions by Exofish.

Similarly, we investigated the amount of overlap between gene annotations in the human GoldenPath and the 76,866 ecores found by Exofish. Genes have been annotated by several methods that include GenomeScan and Ensembl (see methods). There are 42,543 and 35,500 annotations in GenomeScan and Ensembl respectively, and our results show that both sets of predicted genes include a large fraction of ecores, but that a substantial number remain outside annotations (Table 2). According to this analysis, between 33% and 55% of the coding DNA in the genome remains to be annotated by GenomeScan and Ensembl respectively. Conversely, 36.9% and 42.8% of Ensembl and GenomeScan annotations contain ecores, compared to 65% for full-length genes in RefSeq. It is likely that this lower sensitivity on annotations reflects the fragmented nature of the genomic DNA that serves as template for gene predictions (the GoldenPath). Many sequence gaps remain in genomic clones and many sequence contigs cannot yet be oriented. For this reason, genes that span such fragmented regions will often be predicted as several partial genes, with each part containing a fraction of the exons of the whole gene. This may also explain why predicted genes in GenomeScan and Ensembl contain on average a lower number of ecores per genes compared to full-length genes in RefSeq (Table 2).

2.3 Number of Genes in Pufferfish

In the course of the evolution of a genome, new genes are created by duplication of existing genes. In addition to local duplications, a number of whole genome duplications are likely to have occurred in the evolution of eukaryotic lineages. It has been postulated that such a duplication occurred in fish after their divergence from tetrapods (which include mammals) based on a number of results comparing specific genes or gene families between the zebrafish *Danio rerio*, the pufferfish *Fugu rubripes* and human (Amores et al. 1998; Wittbrodt et al. 1998; Aparicio 2000).

The most straightforward argument for vertebrate genome duplications comes from the analysis of gene numbers. By comparing the number of ecores that Exofish identifies respectively in human and in *Tetraodon*, we would expect to obtain a measure of the relative number of genes in each genome. If a duplication has occurred in fish, the third of the *Tetraodon* genome used as a basis for Exofish should contain twice as many ecores as those identified in the human genome sequence available. The reason for this lies in the postulate that although each duplicate copy of a gene in fish will progressively diverge, both should remain partially homologous to their human single copy orthologue.

We found that Exofish identifies 0.8 times the number of ecores in *Tetraodon* compared to human (Table 3). The additional ecores found in human are likely to be due to gene families that are entirely detected by

Table 3. Summary of ecores identified in the *Tetraodon* genome by Exofish comparisons with the *H. sapiens, D. melanogaster* and *C. elegans* genomes

	H. sapiens	D. melanogaster	C. elegans
Number of ecores in the *H. sapiens, D. melanogaster* or *C. elegans* genome in comparison with the 0.4 *Tetraodon* genome equivalents	76,866	9,927	6,581
Number of *Tetraodon* ecores in 0.4 *Tetraodon* genome equivalents in comparison to the *H. sapiens, D. melanogaster* or *C. elegans* genome	64,418	14,088	8,799
Ratio between the number of *Tetraodon* ecores and the number of *H. sapiens, D. melanogaster* or *C. elegans* ecores	0.8	1.4	1.3

only a few members of the orthologous gene family present in the 33% of the *Tetraodon* genome. For instance all ecores corresponding to histones may be detected in human by a single histone present in the available *Tetraodon* sequence. The similarity in ecore numbers between *Tetraodon* and human suggests that pufferfish possess a similar number of genes to human, i.e. approximately 30,000. Our analysis between the entire human genome and a third of *Tetraodon*, in contrast to studies based on specific genes or gene families, provides evidence against a whole genome duplication, at least in pufferfish.

2.4 Ecore Relationships in Metazoans

Conventional views of genome evolution might predict that *Tetraodon* and *C. elegans* will not share genes that are absent in *H. sapiens*. However, we determined that ~5% ([171+301]/8,799) of *Tetraodon* ecores are specifically identified by *C. elegans* (Fig. 1). This result thus reflects either human genes that are not in the GoldenPath and which

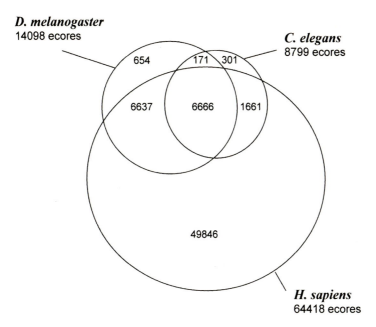

Fig. 1. Relationships between ecores identified in the *Tetraodon* genome by Exofish analysis of the *H. sapiens, D. melanogaster,* and *C. elegans* genomes. When in a region of overlap between two or three circles, numerical values represent the number of ecores that are co-identified by two or three different genomes in *Tetraodon* and counted relative to the smaller genome. When outside a region of overlap, a numerical value represents the number of ecores specifically identified by the genome in *Tetraodon*. For example, a set of 6,837 ecores is co-identified in *Tetraodon* by the three genomes tested, while 49,846 ecores are specifically identified by *H. sapiens* in *Tetraodon*

remain to be sequenced, human genes that have evolved faster than their *Tetraodon* orthologue, or genes that have been lost in the human lineage. Similarly, the 22% ([301+1,661]/8,799) of ecores detected by *C. elegans* that are shared with vertebrates (because they are detected in *Tetraodon*) but not by *D. melanogaster* represent genes that either have not yet been sequenced in the *D. melanogaster* genome, have strongly diverged in the arthropod lineage or have been completely lost.

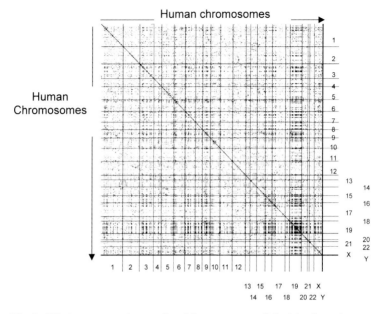

Fig. 2. Whole genome dot matrix of human ecores linked by homology to *Tetraodon* ecores. Human ecores are ordered according to their position in their respective chromosome on both axes. For practical reasons, ecores were grouped in a scanning window of 100 ecores. *Horizontal* and *vertical lines* define the boundaries between chromosomes. Each *black dot* represents the intersection on the X and Y axes of two windows that are homologous to at least one *Tetraodon* ecore in common. The *diagonal* (*top left to bottom right*) is created by ecores being homologous to themselves, and therefore produces the chromosomal axis

2.5 Matrices of Conserved Human Genes

Ecores represent assemblies from several *Tetraodon* sequences alignments over human exons. Accordingly, two or more human genes homologous to each other will contain ecores that include alignments of the same *Tetraodon* sequences, provided the genes are conserved between fish and human. *Tetraodon* ecores may, therefore, be used to relate two or more human genes that are homologous, and these relations may be represented as a dot matrix where human ecores are ordered along their respective chromosomes according to their order in the GoldenPath (Fig. 2). A global view of the human genome appears where coding regions are linked to each other that presumably originate from a common ancestral gene. This representation focuses on conserved genes or gene domains, and, therefore, is not contaminated by low complexity or repeat regions that may create unwanted regions of homology. The matrix has 76,866 co-ordinate points on each axis (one per human ecore) and for practical reasons is schematically represented at a lower resolution in Fig. 2. As a consequence, this representation will only reveal the distribution of large gene clusters, gene families or domains that are multiply represented in specific regions of the human genome. Considering all chromosomes, chromosome 19 stands out with a large number of ecores sharing homologies with other ecores distributed in the genome. We retrieved 135 RefSeq genes that are identified by the *Tetraodon* sequences that are responsible for creating the high density of links on chromosome 19. Out of these, 85% are clearly identified as zinc finger genes, while the remaining 15% do not yet have a description in RefSeq. This result highlights the high density of zinc finger genes on chromosome 19 and clearly indicates the other regions of the human genome that carry zing finger gene clusters, such as chromosomes 1p, 3, 7, 8 and 9 (Fig. 2).

We investigated other regions of the human genome at a much higher resolution, by focusing on homologies between ecores positioned on the same chromosome. Chromosome 17p12-p13 displays a region with an unusual pattern of ecore distribution (Fig. 3A). The dot matrix reveals six tandem repetitions of ~30 ecores, with the sixth repetition on the proximal side being in an inverted orientation. RefSeq genes matched by *Tetraodon* sequences involved in this pattern show that the human ecores correspond to the myosin heavy chain (MyHC) gene cluster. The gene organization of this region has been described in human and mouse (Weiss et al. 1999) as spanning ~350 kb, with all genes being transcribed in the same direction in mouse. In human, the direction of transcription of the proximal gene MyHC-*emb* could not be determined at the time and it was assumed that it is the same as in mouse. Exofish analysis of the region shows that the cluster spans 490 kb and that the human MyHC-*emb* gene, in contrast to the situation in mouse, is in an inverted orientation with respect to the five other MyHC genes. On chromosome 6p21, we examined a 14-Mb region spanning the human

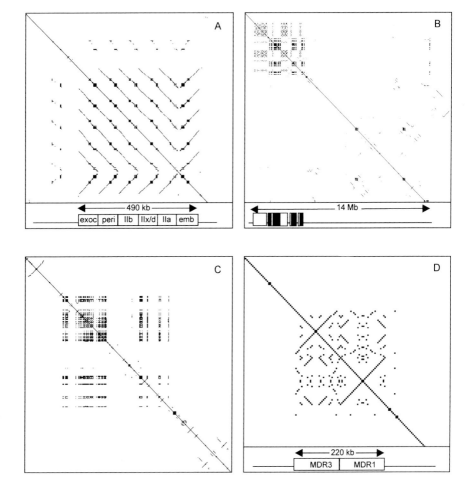

Fig. 3a–d. Legend see p. 23

major histocompatibility complex (MHC) region (Fig. 3B). The first 4.5 Mb contain five histone gene clusters interspersed with four zinc finger gene clusters, while the remaining 9.5 Mb contain ecores with homologies to 428 different human RefSeq genes, of which only five can be described as being involved in the immune response. In view of the clustered nature of MHC genes in vertebrates, it is possible that *Tetraodon* MHC genes are not well represented in the sequence database. However, some categories of genes, especially those involved in the adaptive immune response, may not be conserved enough between human and fish to be identified with Exofish. On chromosome 11p, we examined a region spanning 7.8 Mb containing seven dense clusters of homologous ecores (Fig. 3C). RefSeq genes that correspond to the *Tetraodon* sequences in this region indicate that the ecores belong to the olfactory receptor subfamily. Finally, a 220-kb region in chromosome 7p was studied (Fig. 3D) and shows two successive inverted duplications that reflect the organization of the MDR3 and MDR1 genes of the ATP-binding cassette subfamily. Each gene consists of two halves, in inverted orientation (Chen et al. 1990; Lincke et al. 1991), which are very similar to each other. The distribution of homologous ecores in each gene shows that in addition to the major inversion, exons within each of the halves are also homologous to each other, suggesting that a complex series of rearrangements has given rise the current gene organization. These results provide a possibility of representing relationships

◄ _____   _____

Fig. 3A–D. Dot matrix of ecores in specific regions of the genome (zoomed from the whole genome representation shown in Fig. 2). Here no scanning window was used, therefore *black dots* represent the intersection between two homologous human ecores. The *diagonal* represents the chromosomal axis, where all ecores are homologous to themselves (*top left to bottom right*=pter to qter). **A** A 490-kb region on chromosome 17p12 containing the myosin heavy chain gene clusters (MyHC). The *bottom part* schematizes the organization of the 6 genes in the cluster. **B** A 14-Mb region on chromosome 6p22 shows in the telomeric half 5 histone genes (❑) interspersed with 4 zinc finger genes or gene clusters (■). The large region of 9.5 Mb to the *right of this cluster* spans the major histocompatibility complex (MHC) region and the absence of strong homologous blocks outside the diagonal denotes the near absence of ecores related to MHC genes. **C** A 7.8-Mb region of chromosome 11p11 shows several clusters of ecores homologous to the olfactory receptor subfamily. **D** A 220-kb region of chromosome 7q21 containing the MDR3 and MDR1 genes of the ATP-binding cassette subfamily (see text for details)

between human coding DNA across the entire genome based on homology, with the additional resolution of focusing on single chromosomes
or sub-chromosomal regions. Such representations are specific to genes
conserved among vertebrates and are not obscured by repetitive sequences that would inevitably interfere if the human DNA had not first
been selected through Exofish.

3 Discussion

We have shown that it is possible to compare the DNA of eukaryotic
organisms in order to specifically identify genes. In the case of the
human genome, the identification of ecores should considerably support
the process of annotating the genes in the genome, particularly in view
of our observation that a large fraction of the coding DNA has not yet
been identified by methods based on conventional homology searches
and gene prediction software. On the other hand for *D. melanogaster*
and *C. elegans,* Exofish showed that both genomes have been annotated
to near completion, which raises the question of the reasons for such
differences. As for *H. sapiens*, both the *D. melanogaster* and *C. elegans*
genome have been annotated by a combination of gene predictions and
homology searches. Homologies to cDNAs and spliced ESTs of the
same species are the strongest evidence for the presence of a gene in a
genomic region, and the two model organisms did not benefit from
significantly more cDNA and EST sequence data than *H. sapiens*.
Approximately 39% of annotated genes in *C. elegans* chromosomes and
65% in *D. melanogaster* showed matches to ESTs (Rubin et al. 2000;
Lander et al. 2001). By comparison, complete cDNA sequences exist for
approximately 10,000 out of the 30,000 estimated human genes (33%)
(Maglott et al. 2000), and it has been estimated under stringent criteria
that the 1.6 million human EST sequences available in the public domain represent a "tag" for at least 80% of the genes (Ewing and Green
2000). It is likely that further interactions with human experts for specific gene families and chromosome regions, as was required for annotating the fly and worm genomes, will bring together the vast amount of
knowledge that exists in the biology community to build upon the
current set of predictions and annotate the full complement of human
genes.

We have shown here, with Exofish, glimpses of the evolutionary relationships between four metazoan genomes. Since all mammals share with *Tetraodon* the same common ancestor as do human and *Tetraodon*, it is possible to use Exofish to examine genes in other mammalian species with the same sensitivity and specificity without modifying the parameters that control the selection of alignments. As the sequence of the mouse genome becomes available, together with the completion of the *Tetraodon* genome sequence, it will be of great interest to perform three-way comparisons between *Tetraodon*, human and mouse. Scanning the respective genomes for their specificity in comparison to the common core genetic component should reveal the fundamental features that shape the vertebrate genome.

4 Methods

4.1 Sequencing

The generation and characteristics of *Tetraodon* sequences used for all Exofish comparisons are as described in Roest Crollius et al. (2000a). The generation of the entire sequence of *Tetraodon* is underway at Genoscope.

4.2 Genome Comparisons

Comparisons between the human GoldenPath sequences (July 2000 and September 2000 versions) and *Tetraodon* were computed on the Bio-Cluster at the Compaq Enterprise System Lab (Littleton, Mass., USA). Hardware consisted of 25 compute nodes ES40 with four processors (667 MHz Alpha EV67) and 4 GB of memory each. Fragments of 150 kb of human DNA were compared to the *Tetraodon* database. Exofish makes use of TBLASTX implemented in the LASSAP (large-scale sequence comparison package) software (Glémet and Codani 1997). The comparison between *Tetraodon* and human DNA was completed in 44 h. Comparisons between *Tetraodon* and the *D. melanogaster* and *C. elegans* genomes were computed at Genoscope on a cluster of four Digital Compaq GS60 computers with four processors (525 MHz

Alpha EV6) and 4 GB of memory each and were completed in 32 h and 28 h respectively, on a distributed system used in parallel for other analyses. *Drosophila* gene annotations were retrieved from the Berkeley Drosophila Genome Project (file aa_gadfly.dros) and *C. elegans* annotations were from Wormpep release 20. Ecores identified in the Golden-Path are available at http://www.genoscope.cns.fr/exofish and are positioned on the GoldenPath sequence and displayed on the GenomeBrowser at http://genomebrowser.org/

4.3 GenomeScan and Ensembl

GenomeScan combines the models of exon–intron and splice signal composition used by Genscan (Burge and Karlin 1997) with the similarity information resulting from BLASTX comparison of the query genomic sequence against peptides from the non-redundant protein database in an integrated probabilistic model. The algorithm balances the contributions of these two information sources to infer the most likely gene structure(s) in a genomic sequence. All genes inferred by GenomeScan have at least weak similarity to one or more known proteins. Exon and gene boundaries assigned by GenomeScan are more accurate than those assigned by Genscan when a protein with at least modest similarity to the target gene is available, and the apparent rate of false positives is far lower. Approximately 10% of gene structures inferred by GenomeScan are likely to represent pseudogenes. Ensembl is a joint project between EMBL-EBI and the Sanger Centre (Wellcome Trust Genome Campus, Hinxton, UK) to develop a software system which produces and maintains automatic annotation on eukaryotic genomes. To annotate the sequence of the human genome, exon positions are first identified by analysing raw DNA contigs using Genscan. Predicted exons are then used to search protein, mRNA and EST databases, and high scoring sequences are analysed with GeneWise on the GoldenPath. Genes are built using GeneWise predictions if possible, or with confirmed exons predicted by Genscan.

Acknowledgements. We thank P Wincker, P Brottier, the sequence and template preparation teams, and the computer administration teams at Genoscope; F Francis for critical reading of the manuscript; Compaq Enterprise for access to the Biocluster; Gene-IT for access to the LASSAP software package; the Ensembl project for providing the set of annotated genes on the July 2000 version of the GoldenPath. This work was made possible by the public availability of the *D. melanogaster*, *C. elegans*, and *H. sapiens* genome sequences together with their respective gene annotations, and we thank all contributing sequencing centres and scientists.

References

Adams MD, Celniker SE, Holt RA, Evans CA, Gocayne JD, Amanatides PG, Scherer SE, Li PW, Hoskins RA, Galle RF, George RA, Lewis SE, Richards S, Ashburner M, Henderson SN, Sutton GG, Wortman JR, Yandell MD, Zhang Q, Chen LX, Brandon RC, Rogers YH, Blazej RG, Champe M, Pfeiffer BD, Wan KH, Doyle C, Baxter EG, Helt G, Nelson CR, Gabor Miklos GL, Abril JF, Agbayani A, An HJ, Andrews-Pfannkoch C, Baldwin D, Ballew RM, Basu A, Baxendale J, Bayraktaroglu L, Beasley EM, Beeson KY, Benos PV, Berman BP, Bhandari D, Bolshakov S, Borkova D, Botchan MR, Bouck J, Brokstein P, Brottier P, Burtis KC, Busam DA, Butler H, Cadieu E, Center A, Chandra I, Cherry JM, Cawley S, Dahlke C, Davenport LB, Davies P, de Pablos B, Delcher A, Deng Z, Mays AD, Dew I, Dietz SM, Dodson K, Doup LE, Downes M, Dugan-Rocha S, Dunkov BC, Dunn P, Durbin KJ, Evangelista CC, Ferraz C, Ferriera S, Fleischmann W, Fosler C, Gabrielian AE, Garg NS, Gelbart WM, Glasser K, Glodek A, Gong F, Gorrell JH, Gu Z, Guan P, Harris M, Harris NL, Harvey D, Heiman TJ, Hernandez JR, Houck J, Hostin D, Houston KA, Howland TJ, Wei MH, Ibegwam C et al (2000) The genome sequence of Drosophila melanogaster. Science 287:2185–2195
Amores A, Force A, Yan YL, Joly L, Amemiya C, Fritz A, Ho RK, Langeland J, Prince V, Wang YL, Westerfield M, Ekker M, Postlethwait JH (1998) Zebrafish hox clusters and vertebrate genome evolution. Science 282:1711–1714
Aparicio S (2000) Vertebrate evolution: recent perspectives from fish. Trends Genet 16:54–56
Burge C, Karlin S (1997) Prediction of complete gene structures in human genomic DNA. J Mol Biol 268:78–94
Chen CJ, Clark D, Ueda K, Pastan I, Gottesman MM, Roninson IB (1990) Genomic organization of the human multidrug resistance (MDR1) gene and origin of P-glycoproteins. J Biol Chem 265:506–514
Crnogorac-Jurcevic T, Brown JR, Lehrach H, Schalkwyk LC (1997) Tetraodon fluviatilis, a new puffer fish model for genome studies. Genomics 41:177–184

Ewing B, Green P (2000) Analysis of expressed sequence tags indicates 35,000 human genes. Nat Genet 25:232–234

Fischer C, Ozouf-Costaz C, Roest Crollius H, Dasilva C, Jaillon O, Bouneau L, Bonillo C, Weissenbach J, Bernot A (2000) Karyotype and chromosomal localization of characteristic tandem repeats in the pufferfish *Tetraodon nigroviridis*. Cytogenet Cell Genet 88:50–55

Glémet E, Codani J (1997) Lassap, a large scale sequence comparisons package. CABIOS 13:137–143

Grutzner F, Lutjens G, Rovira C, Barnes DW, Ropers HH, Haaf T (1999) Classical and molecular cytogenetics of the pufferfish Tetraodon nigroviridis. Chromosome Res 7:655–662

Hinegardner R (1968) Evolution of celullar DNA content in Teleost fishes. Am Nat 102:517–523

Lamatsch DK, Steinlein C, Schmid M, Schartl M (2000) Noninvasive determination of genome size and ploidy level in fishes by flow cytometry: detection of triploid Poecilia formosa. Cytometry 39:91–95

Lander ES, Linton LM, Birren B, Nusbaum C, Zody MC, Baldwin J, Devon K, Dewar K, Doyle M, FitzHugh W, Funke R, Gage D, Harris K, Heaford A, Howland J, Kann L, Lehoczky J, LeVine R, McEwan P, McKernan K, Meldrim J, Mesirov JP, Miranda C, Morris W, Naylor J, Raymond C, Rosetti M, Santos R, Sheridan A, Sougnez C, Stange-Thomann N, Stojanovic N, Subramanian A, Wyman D, Rogers J, Sulston J, Ainscough R, Beck S, Bentley D, Burton J, Clee C, Carter N, Coulson A, Deadman R, Deloukas P, Dunham A, Dunham I, Durbin R, French L, Grafham D, Gregory S, Hubbard T, Humphray S, Hunt A, Jones M, Lloyd C, McMurray A, Matthews L, Mercer S, Milne S, Mullikin JC, Mungall A, Plumb R, Ross M, Shownkeen R, Sims S, Waterston RH, Wilson RK, Hillier LW, McPherson JD, Marra MA, Mardis ER, Fulton LA, Chinwalla AT, Pepin KH, Gish WR, Chissoe SL, Wendl MC, Delehaunty KD, Miner TL, Delehaunty A, Kramer JB, Cook LL, Fulton RS, Johnson DL, Minx PJ, Clifton SW, Hawkins T, Branscomb E, Predki P, Richardson P, Wenning S, Slezak T, Doggett N, Cheng JF, Olsen A, Lucas S, Elkin C, Uberbacher E, Frazier M et al (2001) Initial sequencing and analysis of the human genome. Nature 409:860–921

Lincke CR, Smit JJ, van der Velde-Koerts T, Borst P (1991) Structure of the human MDR3 gene and physical mapping of the human MDR locus. J Biol Chem 266:5303–5310

Maglott DR, Katz KS, Sicotte H, Pruitt KD (2000) NCBI's LocusLink and RefSeq. Nucleic Acids Res 28:126–128

Roest Crollius H, Jaillon O, Bernot A, Dasilva C, Bouneau L, Fizames C, Wincker P, Brottier P, Quetier F, Saurin W, Weissenbach J (2000a) Human gene number estimate provided by genome wide analysis using *Tetraodon nigroviridis* genomic DNA. Nat Genet 25:235–238

Roest Crollius H, Jaillon O, Dasilva C, Ozouf-Costaz C, Fizames C, Fischer C, Bouneau L, Billault A, Quetier F, Saurin W, Bernot A, Weissenbach J (2000b) Characterization and repeat analysis of the compact genome of the freshwater pufferfish *Tetraodon nigroviridis*. Genome Res 10:939–949

Rubin GM, Hong L, Brokstein P, Evans-Holm M, Frise E, Stapleton M, Harvey DA (2000) A Drosophila complementary DNA resource. Science 287:2222–2224

Weiss A, McDonough D, Wertman B, Acakpo-Satchivi L, Montgomery K, Kucherlapati R, Leinwand L, Krauter K (1999) Organization of human and mouse skeletal myosin heavy chain gene clusters is highly conserved. Proc Natl Acad Sci USA 96:2958–2963

Wittbrodt J, Meyer A, Schartl M (1998) More genes in fish? Bioessays 20:511–515

3 Lessons from the Sequence of Human Chromosome 22

I. Dunham

1 Introduction

Chromosome 22 is the second smallest of the human autosomes and comprises approximately 1.6%–1.8% of the human genome. It is one of the five human acrocentric chromosomes which each share extensive sequence similarity on the short (p) arm. 22p largely consists of a series of tandem repeat sequence arrays including the ribosomal RNA genes, although there is currently no evidence of any protein-coding genes. In contrast, relative to other chromosomes, the long arm (22q) is rich in protein-coding genes (Deloukas et al. 1998; International Human Genome Sequencing Consortium 2001). The heterochromatic regions comprising the tandem repeats in 22p and the centromere, are very difficult to map and sequence and hence were excluded from the genomic sequencing. Since 22q is a relatively small but discrete unit of the human genome, and had a high-resolution framework map (Collins et al. 1995), it was targeted for early sequence completion. The complete sequence was determined late in 1999 (Dunham et al. 1999b). Here I will describe how the sequence was completed and some observations and developments that have arisen its analysis.

2 The Mapping and Sequencing of Human Chromosome 22

A hierarchical strategy was used to map and sequence chromosome 22q (Fig. 1). Initial long-range maps of chromosome 22 had been established in yeast artificial chromosomes (YAC) (Collins et al. 1995) and by radiation-hybrid mapping (Schuler et al. 1996). However YACs proved to be poor vectors for sequencing human DNA compared to *E. coli*-based cloning vectors, and so additional high-resolution bacterial clone maps were required. Clones representing parts of chromosome 22 were identified by screening extensive bacterial artificial chromosome (BACs) and P1-derived artificial chromosome (PAC) libraries using chromosome 22 sequence tagged site (STS) markers, or by using cosmid and fosmid libraries derived from flow-sorted chromosome 22, and

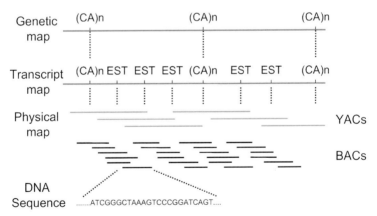

Fig. 1. Hierarchical strategy for mapping and sequencing human chromosome 22. The top-down strategy is described in the text. Each level represents a type of map, with the *tick marks* being genetic (*CA*) or gene-based (*EST*) markers. In the YAC and BAC levels, clones containing DNA corresponding to human chromosome 22 are illustrated by the *horizontal lines* with a minimum sequence tile path picked out by *bolder lines*

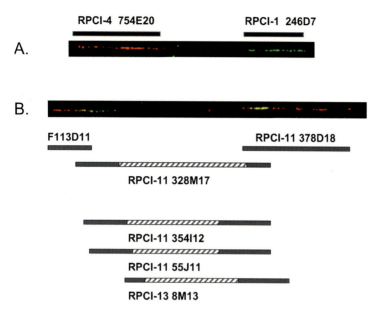

Fig. 2. A An example of DNA fibre FISH data across one of the gaps in the chromosome 22 sequence. The image is oriented centromere to telomere, with the positions of the two clones illustrated beneath. In this fibre, the gap is approximately one clone length. **B** DNA fibre FISH illustrating the presence of a deleted clone that spans a gap. All three BACs from the RPCI-11 library and one from the RPCI-13 library show clear evidence of a deletion on DNA fibres prepared from two different lymphoblastoid cell lines. The approximate composition of the deleted clones is illustrated in the schematic underneath, with the *hashed regions* representing the presumed deletion regions

were assembled in to overlapping clone contigs based on their restriction enzyme fingerprints and STS content. The nascent contigs were ordered relative to each other using the long-range maps. Contigs were extended and joined by cycles of chromosome walking, using sequences obtained from the end of each contig (see Dunham et al. 1999a for methods). In two places, YAC clones had to be used to join or extend contigs. After this stage, 22q was contained in 11 bacterial clone contigs stretching from centromere to telomere. The largest contig covered more than 23 Mb. Most of the gaps between contigs were sized using either

DNA fibre fluorescence in situ hybridisation (FISH) or restriction enzyme mapping and were established to be less than 150 kb. No further clones could be identified for the gap regions despite screening many libraries. For three gaps, BAC and PAC clones that contain STSs on either side of the gap were shown by DNA fibre FISH to be deleted for at least a minimal core region (Fig. 2A). In one case, further analysis showed that despite identifying four BAC clones from two different library sources, each of the clones was deleted for the gap region. Furthermore, the deletion was consistently seen in DNA fibres from two different lymphoblastoid cell lines, indicating that it was unlikely to be a natural polymorphism (Fig. 2B). Hence, we speculated that the gaps represent regions of the human genome that are unclonable in current *E. coli* vector systems (Dunham et al. 1999b).

Since publication, Bruce Roe's group at the University of Oklahoma, with the help of collaborators, have identified and sequenced three BACs which closed the third gap from the centromere in 22q11 (B. Roe, personal communication). Analysis of the sequence in this gap identified three genes. We also have evidence of a pair of genes with 3' ends present in the finished sequence, but whose 5' ends extend into one of the 22q13 uncloned sequence gaps. Clearly there is interesting sequence to be determined in the gaps and work continues to try to complete the regions.

3 Genomic Sequencing

Each of the four sequencing groups involved took responsibility for completion of the sequence in defined areas of 22q as illustrated in Fig. 3. Genomic sequence was determined by random shotgun sequencing in M13 and pUC vectors for each of a representative set of minimally overlapping clones chosen to be sufficient to completely cover the physical map (the "tile path"). At the Sanger Centre, the shotgun sequence was assembled using the Phrap assembly algorithm (Ewing et al. 1998; Ewing and Green 1998) and the assembly imported in to a Genome Assembly Program (GAP) database (Bonfield et al. 1995) for directed analysis and sequencing to close gaps and resolve ambiguities ("finishing"). The major problems encountered during the directed sequencing phase were due to CpG islands, tandem repeats and apparent

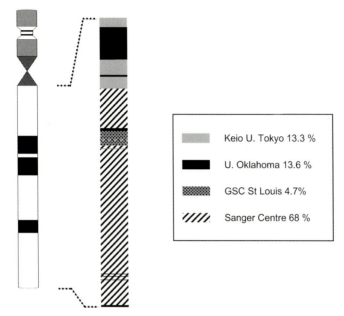

Fig. 3. Division of sequencing between the four sequencing groups for human chromosome 22q. For full details of the sequencing groups see Dunham et al. 1999b

cloning biases. Directed sequencing using oligonucleotide primers (see for instance, Wilson and Mardis 1999), very short insert plasmid libraries (McMurray et al. 1998) (Fig. 4), or identification of bridging clones by screening high-complexity plasmid or M13 libraries (Flint et al. 1998) solved these problems (Table 1).

In the final sequence, one additional gap close to the centromere was intractable to sequencing. Hence the complete sequence of chromosome 22q consisted of 12 contigs covering 33.4 Mb estimated to be accurate to less than 1 error in 50,000 bases (Dunham et al. 1999b). The gaps were localised to the two ends of the map, separated by the largest contiguous segment of 23 Mb. Taking into account the gap sizes, it was estimated that the sequence was complete to 97% coverage of 22q (Table 2).

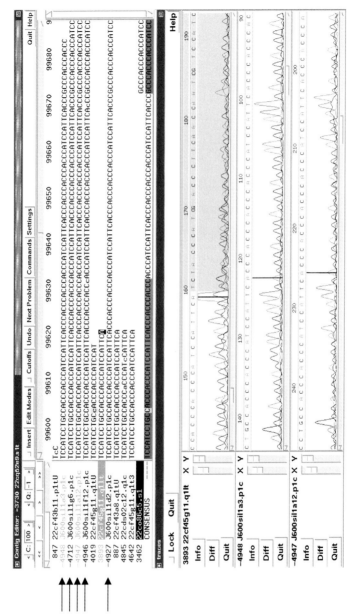

Fig. 4. Short insert library (SIL) over imperfect tandem repeat provides new sequence resulting in good-quality sequence. The *arrowed* sequences are derived from a SIL cloned from a restriction fragment covering what would otherwise be a gap in the sequence. Only these SIL sequence cover the central region

Table 1. Types of problematical sequence on human chromosome 22 and methods for solving these problems

Problem	Sanger solutions
GC-rich regions/CGI	SIL
	M13 libraries
Alu at contig ends	SIL
Tandem repeats	PCR, SIL
Cloning bias	PCR, Oligo screening of M13 library

SIL, short insert library(McMurray et al. 1998); CGI, CpG island sequence.

Table 2. Chromosome 22 sequence contigs, including gap closure from the University of Oklahoma group

Contig	Size (kb)
AP000522–AP000529	234
AP000530–AP000542	406
AP000543–AC007663	3,457
AC007731–AL049708	23,006
AL118498–AL022339	767
Z78594–AL049811	1,528
AL049853–AL096853	2,485
AL096843–AL078607	190
AL078613–AL117328	993
AL080240–AL022328	291
AL096767–AC002055	380
Total sequence length	33,737
Total length of 22q	34,663

Contigs are listed by the first and last database accession number in centromere to telomere orientation.

4 Sequence Analysis and Gene Content

The approach taken to identify genes in the chromosome 22 sequence relied on a combination of sequence similarity searches with expressed sequences and ab initio gene predictions. Interspersed and tandem repeats in the sequence were identified and masked. A series of basic local alignment search tool (BLAST) similarity searches (Altschul et al. 1990) were used to compare the masked sequence to public domain DNA and protein databases. Gene prediction software was used to analyse the contiguous masked sequence (Burge and Karlin 1997; Solovyev and Salamov 1997). The analysis was imported in to an ACEDB database (Durbin and Thierry-Mieg 1991) and gene features were manually annotated. In total 545 gene and 134 pseudogene sequences were annotated across 22q (Dunham et al. 1999b). In addition 118 immunoglobulin lambda variable gene segments were present but these were counted as components of a single locus in the annotation statistics. The gene annotations were divided into three categories by the perceived strength of their supporting evidence. These were (1) those identical to a known human gene or protein sequences (n=247), (2) those similar to, or containing a region of similarity to, any other gene or protein (n=150), or (3) those supported only by expressed sequence tags (ESTs) (n=148). The pseudogenes are predominantly (82%) processed pseudogenes lacking the characteristic intron–exon structure of the putative parent gene, with the others being non-functional segments of duplicated genes. Another 325 ab initio gene predictions were produced by Genscan that did not overlap with the annotations supported by expression data. Based on previous experiments that calibrated the rate of over-prediction by Genscan, it was estimated that perhaps 1/3 of these might represent real genes.

Since publication, new sequence data from the public databases and experimental analysis annotation has confirmed that some of the original annotations represented multiple fragments of single genes, so that these were fused into revised structures. The annotation has been extended to identify 17 new genes or pseudogenes. The extended annotation has also added 440 exons (12% increase) on to the previous structures. In total, this work represents 133.8 kb of additional annotation giving a total of 1.43 Mb of "cDNA" equivalent length for chromosome 22q. Thus, there are now 551 genes and 141 pseudogenes annotated on

Table 3. Estimates of the number of genes in the human genome based on the annotation of the finished sequences of chromosomes 21 and 22

Data	Genome fraction	Genes (predictions)	Estimate	Corrected (transcript 1999)
Chr 22	0.011	556	50,100	36,300
		(651)	58,650	42,500
Chr 21	0.011	193	17,200	21,000
		(225)	20,100	24,500
Both	0.022	749	33,500	30,500
		(876)	39,000	35,750
Mean				35,000

The numbers in brackets in the genes column represent an estimate of the total number of gene including ab initio predictions unsupported by expression data. The estimates are simply based on scaling the number of genes on chromosome 22 or 21 or both to a presumed genome size of 3 Gb. In the correction column the estimates are corrected to take account of observed differences in gene density compared to the rest of the genome as described in the transcript map 1999 (Deloukas et al. 1998).

chromosome 22, excluding the gene segments of the immunoglobulin lambda variable locus (Dunham 2000). The new chromosome 22 annotation can be combined with the annotation provided from the recently finished genomic sequence of chromosome 21(Hattori et al. 2000), a relatively gene-poor chromosome, to provide an estimate of the total number of genes that are likely to be found in human. Together, chromosome 21 and 22 account for 2% of the genome and under a variety of assumptions produce a mean estimate of approximately 35,000 genes in the genome (Table 3).

Overall, the chromosome 22 sequence occupied by annotated genes, including their introns, is about 39% of the total sequence. However, only about 4% of this is used for exons. The size of individual genes encoded on chromosome 22 varies dramatically from approximately 1 kb in length to over 583 kb (Fig. 5). Two cases of genes occurring within the introns of other expressed genes were observed, although the nested genes were in the opposite transcriptional orientation in both cases. One of the nested gene structures provides a fascinating glimpse of the evolutionary stability of this type of structure. The tissue inhibitor

Smallest gene:

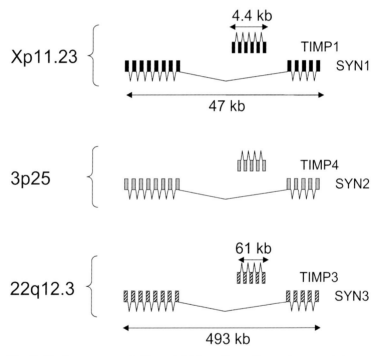

Fig. 5. The extremes of gene size on chromosome 22. The *boxes* represent exons with the *filled regions* containing the coding open reading frame and the *hashed regions* being the untranslated regions

Fig. 6. The human synapsin/tissue inhibitor of metalloproteinase nested gene family. For each of the three loci, the TIMP gene is on the *top strand* in 5'-3' orientation left to right, while the synapsin gene is on the *bottom strand* in the opposite orientation

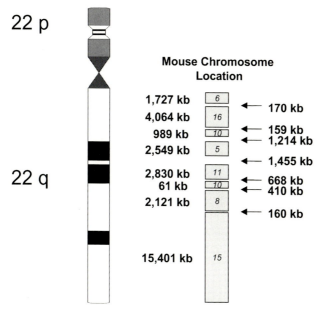

Fig. 7. Conservation of synteny between human chromosome 22 and regions of the mouse genome. The corresponding mouse chromosomal regions as identified by the positions of genes orthologous to those on human chromosome 22, are indicated by the *numbered boxes*. The sizes of the region on human chromosome 22 are given in kilobases on the *left*, while the sizes of the regions that must contain the breakpoints of conserved synteny are given to the *right*

of metalloprotease (TIMP)3/synapsin (SYN)3 nested gene structure has two duplicated copies elsewhere in the human genome, on chromosomes Xp11.23 and 3p25. Examination of the structure of these copies in finished and working draft genomic sequence indicates a common evolutionary origin with each of the TIMP genes being buried in the fifth intron of the corresponding synapsin gene (Fig. 6). Although the intron–exon organisation of these nested pairs is remarkably similar, there is a tenfold difference in the genomic size between the TIMP1/SYN1 genes on the X chromosome and the TIMP3/SYN3 genes on chromosome 22. Previous work suggests that at least the X chromosome nested gene pair is also present in mouse (Derry and Barnard 1992). Remarkably, in *Droso-*

phila the same nested gene structure is also found (Pohar et al. 1999) (see also the genomic sequence Accession: AE003686; Adams et al. 2000). This nested gene structure does not appear to occur in the genome of the nematode *C. elegans,* and it will be interesting to trace this fossil gene structure through the genomes of many more organisms to obtain a complete picture of when and perhaps how it arose.

Comparison of the position and order of the human chromosome 22 genes to the positions of their mouse orthologues confirmed that human chromosome 22 is represented by eight conserved linkage groups in the mouse, on mouse chromosomes 6, 16, 10, 5, 11, 8 and 15 (Fig. 7). Since completion of the sequence, one of the corresponding synteny break regions has been sequenced in the mouse (Pletcher et al. 2000). Further sequencing in the mouse of the other regions where conserved synteny breaks down should shed light on the processes that have been involved in the genome shuffling that has occurred since human and mouse shared a common ancestral genome.

5 DNA Sequence Variation

The DNA sequence of different individuals varies in a number of ways. Historically, the length polymorphisms of tandem repeats (so-called mini- and micro-satellites) have been a very useful source of DNA markers for genetic mapping. Similarly, insertion/deletion differences have been identified between different individuals, some of which involve genes. However, there is another type of variation which is more common and potentially of great value in future genetics, whether studies of Mendelian diseases (Kruglyak 1997), population histories and genetics (Chakravarti 1999), or personalisation of medicine (Housman and Ledley 1998). This is the single nucleotide polymorphism (SNP). SNPs have high abundance in the genome, occurring at ~1 per kb when any two human genomes are compared. Furthermore, the biallelic nature of SNPs should prove to be amenable to high throughput automated genotyping using a number of new technologies.

The availability of large amounts of finished chromosome 22 genomic sequence has provided an opportunity to pilot methods to identify many of these genetic differences (Dawson et al. 2001; Mullikin et al. 2000). In the first approach, we identified SNPs and small

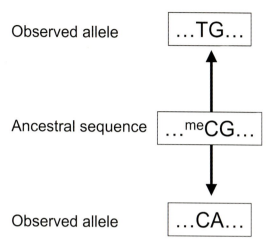

Fig. 8. Possible mechanism for the origin of CA↔TG type adjacent SNPs. An ancestral methylated CpG dinucleotide could first mutate to either TG or CA to form one of the new alleles. A second independent event could then occur on another individual with the ancestral methylated CpG to give the reciprocal allele. This process could give rise to a biallelic CA/TG system in the CpG is then lost by genetic drift, or a population bottleneck. Alternatively, a multi-allelic system may be fixed in the population

insertion/deletion polymorphisms from the overlapping sections of bacterial clones in the sequence tile path, which represent different sequence haplotypes. This was possible because the set of clones used to determine the sequence of chromosome 22 came from multiple clone libraries derived from the DNA of different individuals. Whenever neighbouring clones in the tile path come from libraries from different individuals and on average 50% of the time when they are form the same diploid library, the sequence will be derived from different haplotypes and hence will contain SNPs. Using this approach, we were able to identify 12,267 candidate variations across the chromosome in clusters with a mean density of 1 per 1.07 kb (Dawson et al. 2001). Eighty-two percent were SNPs and the remainder small insertion/deletion differences. Experimental verification of a small sample indicated that 92% of the candidate variations were present in a panel of 92 individuals. In the second approach, additional haplotypes represented in new small insert libraries were sequenced as part of the SNP Consortium (http://snp.cshl.org/) project and overlaid onto the chromosome 22 genomic sequence (Mullikin et al. 2000). SNPs were identified by sensitive comparison criteria to produce a collection of 2,730 randomly distributed SNPs. Taken together, there is now an extensive resource of more than 15,000 chromosome 22 SNPs and other small variations. This resource of SNPs will be invaluable in future genetic studies of chromosome 22, such as investigating the extent of linkage disequilibrium across the chromosome.

The characteristics of these SNPs confirmed the observations of previous smaller studies. Approximately 70% were transition base changes (i.e. A↔T or G↔C). For the SNPs detected from overlaps we also identified a tenfold increase over the expected rate of SNP changes occurring at adjacent bases (Dawson et al. 2001). Analysis of the frequencies of the specific base changes for these "double SNPs" showed an excess of the type CA↔TG. It is possible that these represent the result of two independent methylated CpG dinucleotide deaminations to give two new alleles as illustrated in Fig. 8.

6 Conclusion

Since completion, the finished chromosome 22 sequence has served as a valuable model for studies of gene and promoter prediction (Ewing and Green 2000; Penn et al. 2000; Roest Crollius et al. 2000; Scherf et al. 2001; Shoemaker et al. 2001), investigation of chromosomal deletions, translocations and duplications in disease (Bruder et al. 2001; Edelmann et al. 2001a,b), SNP identification (Dawson et al. 2001; Mullikin et al. 2000), analysis of linkage disequilibrium (Dunning et al. 2000), recombination rates (Majewski and Ott 2000), production of minichromosomes (Yoshimi et al. 2000) and validation of new EST or cDNA collections (de Souza et al. 2000; Wiemann et al. 2001). Several genes involved in human genetic disease have been identified with the help of the genomic sequence, notably the non-syndromic hereditary deafness DFNA17 (Lalwani et al. 2000), May-Hegglin anomaly (Kelley et al. 2000; May-Hegglin/Fechtner Syndrome Consortium 2000; Seri et al. 2000) and spinocerebellar ataxia 10 (Matsuura et al. 2000). This is testament to the power of finished genomic sequence as both a research tool and a stimulus for new research avenues.

The success of the chromosome 22 sequencing project clearly demonstrates that the clone-by-clone strategy is capable of generating long-range continuity over the genome. Although a collection of bacterial cloning vectors was used to establish the clone maps of human chromosome 22, BACs and PACs will be the predominant cloning vectors for the rest of the genome because of the advantage given by the larger insert size. However, I believe that other vectors will continue to have a role to play, especially in dealing with the problematical gap regions. For chromosome 22 we were unable to obtain sequence over 11 small gaps initially. Although walking in BAC libraries later closed one of these gaps, the remaining gaps have remained. Chromosome walking using fosmids with their smaller insert sizes and low copy number could enable cloning of DNA into these gaps, because putative "unclonable" sequences might be avoided in some of the walking steps. This might be more aptly termed chromosome crawling, edging up to the unclonable regions a step at a time. It is probable that the unclonable sequences are unlikely to be specific to chromosome 22. Hence, the genome sequence could still harbour a handful of troublesome gaps for a number of years beyond "completion".

The initial annotation of chromosome 22 made extensive use of similarity to expressed sequence in the form of cDNA or ESTs but was incomplete because of the inadequacy of these resources. Since publication, there has been a considerable increase in the quantity and quality of publicly available expressed sequence resources, and this has enabled the annotation to be considerably refined, together with new data types such as promoter predictions (Scherf et al. 2001) and *Tetraodon nigroviridis* sequence (Roest Crollius et al. 2000). The future availability of genomic DNA sequence from mouse and a number of other vertebrates should also greatly aid gene annotation. Remarkably, the overall number of genes annotated in the chromosome 22 sequence has not increased very much because although a few new genes have been found, other previous partial annotations have been fused together. At the present time, I do not anticipate that the overall number of genes will increase by more than a few tens.

Acknowledgements. Many thanks are due to all the people who have been involved with the chromosome 22 projects, both from the Sanger Centre and from the consortium of labs worldwide. I would particularly like to thank all the members of the Sanger Centre Chromosome 22 mapping and sequence groups who have made much of this work possible, and members of the Sanger Centre Molecular Cytogenetics group, Adrienne Hunt and David Beare for kindly providing images for some of the figures. The author is supported by the Wellcome Trust.

References

Adams MD, Celniker SE, Holt RA, Evans CA, Gocayne JD, Amanatides PG, Scherer SE, Li PW, Hoskins RA, Galle RF, George RA, Lewis SE, Richards S, Ashburner M, Henderson SN, Sutton GG, Wortman JR, Yandell MD, Zhang Q, Chen LX, Brandon RC, Rogers YH, Blazej RG, Champe M, Pfeiffer BD, Wan KH, Doyle C, Baxter EG, Helt G, Nelson CR, Gabor GL, Abril JF, Agbayani A, An HJ, Andrews-Pfannkoch C, Baldwin D, Ballew RM, Basu A, Baxendale J, Bayraktaroglu L, Beasley EM, Beeson KY, Benos PV, Berman BP, Bhandari D, Bolshakov S, Borkova D, Botchan MR, Bouck J et al (2000) The genome sequence of Drosophila melanogaster. Science 287:2185–2195

Altschul SF, Gish W, Miller W, Myers EW, Lipman DJ (1990) Basic local alignment search tool. J Mol Biol 215:403–410

Bonfield JK, Smith K, Staden R(1995) A new DNA sequence assembly program. Nucleic Acids Res 23:4992–4999

Bruder CE, Hirvela C, Tapia-Paez II, Fransson II, Segraves R, Hamilton G, Zhang XX, Evans DG, Wallace AJ, Baser ME, Zucman-Rossi J, Hergersberg M, Boltshauser E, Papi L, Rouleau GA, Poptodorov G, Jordanova A, Rask-Andersen H, Kluwe L, Mautner VV, Sainio M, Hung G, Mathiesen T, Moller C, Pulst SM, Harder H, Heiberg A, Honda M, Niimura M, Sahlen S, Blennow E, Albertson DG, Pinkel D, Dumanski JP (2001) High resolution deletion analysis of constitutional DNA from neurofibromatosis type 2 (NF2) patients using microarray-CGH. Hum Mol Genet 10:271–282

Burge C, Karlin S (1997) Prediction of complete gene structures in human genomic DNA. J Mol Biol 268:78–94

Chakravarti A (1999) Population genetics – making sense out of sequence. Nat Genet 21:56–60

Collins JE, Cole CG, Smink LJ, Garrett CL, Leversha MA, Soderlund CA, Maslen GL, Everett LA, Rice KM, Coffey AJ et al (1995) A high-density YAC contig map of human chromosome 22. Nature 377:367–379

Dawson E, Chen Y, Hunt S, Smink LJ, Hunt A, Rice K, Livingston S, Bumpstead S, Bruskiewich R, Sham P, Ganske R, Adams M, Kawasaki K, Shimizu N, Minoshima S, Roe B, Bentley D, Dunham I (2001) A SNP resource for human chromosome 22: extracting dense clusters of SNPs from the genomic sequence. Genome Res 11:170–178

de Souza SJ, Camargo AA, Briones MR, Costa FF, Nagai MA, Verjovski-Almeida S, Zago MA, Andrade LE, Carrer H, El-Dorry HF, Espreafico EM, Habr-Gama A, Giannella-Neto D, Goldman GH, Gruber A, Hackel C, Kimura ET, Maciel RM, Marie SK, Martins EA, Nobrega MP, Paco-Larson ML, Pardini MI, Pereira GG, Pesquero JB, Rodrigues V, Rogatto SR, da Silva ID, Sogayar MC, de Fatima Sonati M, Tajara EH, Valentini SR, Acencio M, Alberto FL, Amaral ME, Aneas I, Bengtson MH, Carraro DM, Carvalho AF, Carvalho LH, Cerutti JM, Correa ML, Costa MC, Curcio C, Gushiken T, Ho PL, Kimura E, Leite LC, Maia G, Majumder P, Marins M, Matsukuma A, Melo AS, Mestriner CA, Miracca EC, Miranda DC, Nascimento AN, Nobrega FG, Ojopi EP, Pandolfi JR, Pessoa LG, Rahal P, Rainho CA, da Ros N, de Sa RG, Sales MM, da Silva NP, Silva TC, da Silva W, Jr, Simao DF, Sousa JF, Stecconi D, Tsukumo F, Valente V, Zalcbeg H, Brentani RR, Reis FL, Dias-Neto E, Simpson AJ (2000) Identification of human chromosome 22 transcribed sequences with ORF expressed sequence tags. Proc Natl Acad Sci USA 97:12690–12693

Deloukas P, Schuler GD, Gyapay G, Beasley EM, Soderlund C, Rodriguez-Tome P, Hui L, Matise TC, McKusick KB, Beckmann JS, Bentolila S, Bihoreau M, Birren BB, Browne J, Butler A, Castle AB, Chiannilkulchai N, Clee C, Day PJ, Dehejia A, Dibling T, Drouot N, Duprat S, Fizames C et al (1998) A physical map of 30,000 human genes. Science 282:744–746

Derry JM, Barnard PJ (1992) Physical linkage of the A-raf-1, properdin, synapsin I, and TIMP genes on the human and mouse X chromosomes. Genomics 12:632–638

Dunham I (2000) The gene guessing game. Yeast 17:218–224

Dunham I, Dewar K, Kim U-J, Ross MT (1999a) Cloning systems, vol 3. In: Birren B, Green ED, Klapholz S, Myers RM, Riethman H, Roskams J (eds) Genome analysis: a laboratory manual series, vol 3. Cold Spring Harbor Laboratory Press, Cold Spring Harbor, New York, pp 1–86

Dunham I, Hunt AR, Collins JE, Bruskiewich R, Beare DM, Clamp M, Smink LJ, Ainscough R, Almeida JP, Babbage A, Bagguley C, Bailey J, Barlow K, Bates KN, Beasley O, Bird CP, Blakey S, Bridgeman AM, Buck D, Burgess J, Burrill WD et al (1999b) The DNA sequence of human chromosome 22. Nature 402:489–495

Dunning AM, Durocher F, Healey CS, Teare MD, McBride SE, Carlomagno F, Xu CF, Dawson E, Rhodes S, Ueda S, Lai E, Luben RN, Van Rensburg EJ, Mannermaa A, Kataja V, Rennart G, Dunham I, Purvis I, Easton D, Ponder

BA (2000) The extent of linkage disequilibrium in four populations with distinct demographic histories. Am J Hum Genet 67:1544–1554

Durbin R, Thierry-Mieg J (1991) A C. elegans database. http://www.sanger.ac.uk/Software/Acedb

Edelmann L, Spiteri E, Koren K, Pulijaal V, Bialer MG, Shanske A, Goldberg R, Morrow BE (2001a) AT-rich palindromes mediate the constitutional t(11;22) translocation. Am J Hum Genet 68:1–13

Edelmann L, Stankiewicz P, Spiteri E, Pandita RK, Shaffer L, Lupski J, Morrow BE (2001b) Two functional copies of the DGCR6 gene are present on human chromosome 22q11 due to a duplication of an ancestral locus. Genome Res 11:208–217

Ewing B, Green P (1998) Base-calling of automated sequencer traces using phred. II. Error probabilities. Genome Res 8:186–194

Ewing B, Green P (2000) Analysis of expressed sequence tags indicates 35,000 human genes. Nat Genet 25:232–234

Ewing B, Hillier L, Wendl MC, Green P (1998) Base-calling of automated sequencer traces using phred. I. Accuracy assessment. Genome Res 8:175–185

Flint J, Sims M, Clark K, Staden R, Thomas K (1998) An oligo-screening strategy to fill gaps found during shotgun sequencing projects. DNA Seq 8:241–245

Hattori M, Fujiyama A, Taylor TD, Watanabe H, Yada T, Park HS, Toyoda A, Ishii K, Totoki Y, Choi DK, Soeda E, Ohki M, Takagi T, Sakaki Y, Taudien S, Blechschmidt K, Polley A, Menzel U, Delabar J, Kumpf K, Lehmann R, Patterson D, Reichwald K, Rump A, Schillhabel M, Schudy A (2000) The DNA sequence of human chromosome 21. The chromosome 21 mapping and sequencing consortium. Nature 405:311–319

Housman D, Ledley FD (1998) Why pharmacogenomics? Why now? Nat Biotechnol 16:492–493

International Human Genome Sequencing Consortium (2001) Initial sequencing and analysis of the human genome. Nature 409:860–921

Kelley MJ, Jawien W, Ortel TL, Korczak JF (2000) Mutation of MYH9, encoding non-muscle myosin heavy chain A, in May-Hegglin anomaly. Nat Genet 26:106–108

Kruglyak L (1997) The use of a genetic map of biallelic markers in linkage studies. Nat Genet 17:21–24

Lalwani AK, Goldstein JA, Kelley MJ, Luxford W, Castelein CM, Mhatre AN (2000) Human nonsyndromic hereditary deafness DFNA17 is due to a mutation in nonmuscle myosin MYH9. Am J Hum Genet 67:1121–1128

Majewski J, Ott J (2000) GT repeats are associated with recombination on human chromosome 22. Genome Res 10:1108–1114

Matsuura T, Yamagata T, Burgess DL, Rasmussen A, Grewal RP, Watase K, Khajavi M, McCall AE, Davis CF, Zu L, Achari M, Pulst SM, Alonso E, Noebels JL, Nelson DL, Zoghbi HY, Ashizawa T (2000) Large expansion of the ATTCT pentanucleotide repeat in spinocerebellar ataxia type 10. Nat Genet 26:191–194

May-Hegglin/Fechtner Syndrome Consortium (2000) Mutations in MYH9 result in the May-Hegglin anomaly, and Fechtner and Sebastian syndromes. Nat Genet 26:103–105

McMurray AA, Sulston JE, Quail MA (1998) Short-insert libraries as a method of problem solving in genome sequencing. Genome Res 8:562–566

Mullikin JC, Hunt SE, Cole CG, Mortimore BJ, Rice CM, Burton J, Matthews LH, Pavitt R, Plumb RW, Sims SK, Ainscough RM, Attwood J, Bailey JM, Barlow K, Bruskiewich RM, Butcher PN, Carter NP, Chen Y, Clee CM, Coggill PC, Davies J, Davies RM, Dawson E, Francis MD, Joy AA, Lamble RG, Langford CF, Macarthy J, Mall V, Moreland A, Overton-Larty EK, Ross MT, Smith LC, Steward CA, Sulston JE, Tinsley EJ, Turney KJ, Willey DL, Wilson GD, McMurray AA, Dunham I, Rogers J, Bentley DR (2000) An SNP map of human chromosome 22. Nature 407:516–520

Penn SG, Rank DR, Hanzel DK, Barker DL (2000) Mining the human genome using microarrays of open reading frames. Nat Genet 26:315–318

Pletcher MT, Roe BA, Chen F, Do T, Do A, Malaj E, Reeves RH (2000) Chromosome evolution: the junction of mammalian chromosomes in the formation of mouse chromosome 10. Genome Res 10:1463–1467

Pohar N, Godenschwege TA, Buchner E (1999) Invertebrate tissue inhibitor of metalloproteinase: structure and nested gene organization within the synapsin locus is conserved from Drosophila to human. Genomics 57:293–296

Roest Crollius H, Jaillon O, Bernot A, Dasilva C, Bouneau L, Fischer C, Fizames C, Wincker P, Brottier P, Quetier F, Saurin W, Weissenbach J (2000) Estimate of human gene number provided by genome-wide analysis using Tetraodon nigroviridis DNA sequence. Nat Genet 25:235–238

Scherf M, Klingenhoff A, Frech K, Quandt K, Schneider R, Grote K, Frisch M, Gailus-Durner VV, Seidel A, Brack-Werner R, Werner T (2001) First pass annotation of promoters on human chromosome 22. Genome Res 11:333–340

Schuler GD, Boguski MS, Stewart EA, Stein LD, Gyapay G, Rice K, White RE, Rodriguez-Tome P, Aggarwal A, Bajorek E, Bentolila S, Birren BB, Butler A, Castle AB, Chiannilkulchai N, Chu A, Clee C, Cowles S, Day PJ, Dibling T, Drouot N, Dunham I, Duprat S, East C, Hudson TJ et al (1996) A gene map of the human genome. Science 274:540–546

Seri M, Cusano R, Gangarossa S, Caridi G, Bordo D, Lo Nigro C, Ghiggeri G, Ravazzolo R, Savino M, Del Vecchio M, d'Apolito M, Iolascon A, Zelante LL, Savoia A, Balduini CL, Noris P, Magrini U, Belletti S (2000) Mutations in MYH9 result in the May-Hegglin anomaly, and Fechtner and Sebastian syndromes. Nature Genet 26:103–105

Shoemaker DD, Schadt EE, Armour CD, He YD, Garrett-Engele P, McDonagh PD, Loerch PM, Leonardson A, Lum PY, Cavet G, Wu LF, Altschuler SJ, Edwards S, King J, Tsang JS, Schimmack G, Schelter JM, Koch J, Ziman M, Marton MJ, Li B, Cundiff P, Ward T, Castle J, Krolewski M, Meyer MR, Mao M, Burchard J, Kidd MJ, Dai H, Phillips JW, Linsley PS, Stoughton R, Scherer S, Boguski MS (2001) Experimental annotation of the human genome using microarray technology. Nature 409:922–927

Solovyev V, Salamov A (1997) The Gene-Finder computer tools for analysis of human and model organisms genome sequences. Ismb 5:294–302

Wiemann S, Weil B, Wellenreuther R, Gassenhuber J, Glassl S, Ansorge W, Bocher M, Blocker H, Bauersachs S, Blum H, Lauber J, Dusterhoft A, Beyer A, Kohrer K, Strack N, Mewes HW, Ottenwalder B, Obermaier B, Tampe J, Heubner D, Wambutt R, Korn B, Klein M, Poustka A (2001) Toward a catalog of human genes and proteins: sequencing and analysis of 500 novel complete protein coding human cDNAs. Genome Res 11:422–435

Wilson RK, Mardis ER (1999) Analyzing DNA, vol 1. In: Birren B, Green ED, Klapholz S, Myers RM, Roskams J (eds) Genome analysis: a laboratory manual series, vol 1. Cold Spring Harbor Laboratory Press, Cold Spring Harbor, New York, pp 397–454

Yoshimi K, Tomizuka K, Shinohara T, Kazuki Y, Yoshida H, Ohguma A, Yamamoto T, Tanaka S, Oshimura M, Ishida I (2000) Manipulation of human minichromosomes to carry greater than megabase-sized chromosome inserts. Nat Biotechnol 18:1086–1090

4 ATM: From Phenotype to Functional Genomics – And Back

Y. Shiloh

1 Genetic Diseases Highlight Physiological Functions

The first phase of the Human Genome Project is close to completion. Soon, a complete draft of the vast human genome sequence will be available for analysis. However, the proportion of this phase in the momentous attempt of man to understand human biology is small compared to what remains to be done: to understand the functions of all gene products, their relationships to each other, and the genomic variability that underlies human phenotypic variation. This effort, which may take much longer than the initial sequencing of the human genome, has brought us back to cell biology and biochemistry, albeit with new methods, better equipment, and novel, high throughput strategies like the microarray technology.

    The post-genomic era is essentially the era of functional genomics – the search for the functions of numerous novel sequences, many of which are not annotated and are functionally anonymous. Under-

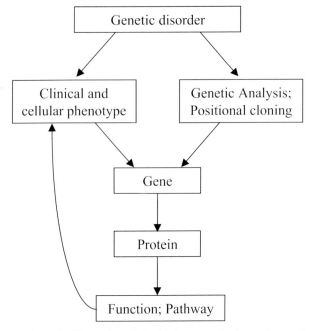

Fig. 1. A schematic illustration of the basic stages in the understanding of a genetic disorder at the molecular and physiological levels

standing their functions is a challenging task, but when a gene is initially associated with a specific phenotype, e.g., a genetic disorder, such a phenotype provides functional insights that can serve as guidelines in further research. A genetic disease usually attracts the attention of clinicians and researchers, with the inevitable cloning of the responsible gene followed by functional analysis of its product. In many cases, this path leads to the discovery of a new physiological function (Fig. 1). It is not surprising, therefore, that genes associated with genetic disorders constitute a major portion of the genes whose functions are already understood. In many cases, particularly when the identification of the disease gene is carried out using positional cloning, the work on a specific disorder involves the entire scope of Human Genome Project tools, on a small scale: genetic and physical mapping, extensive cloning, gene identification, sequence analysis, and finally, the move back to biology and biochemistry. The human genetic disorder ataxia-telangiectasia (A-T) is a prime example of this paradigm. The discovery of the gene responsible for this disease, *ATM*, and its protein product recently revealed a central regulatory junction that controls a striking array of cellular signaling pathways. Due to the multisystem nature of A-T, its investigation has widespread implications for many areas of medicine.

2 Ataxia-Telangiectasia: From Phenotype to Gene

Ataxia-telangiectasia (A-T) is an autosomal recessive disorder with the following characteristics: progressive degeneration of the cerebellar cortex expressed primarily as gradual loss of the Purkinje cells, leading initially to ataxia (lack of balance) and later to general motor dysfunction; oculocutaneous telangiectasias (dilated blood vessels); immunodeficiency spanning the cellular and humoral arms, with recurrent sinopulmonary infections in some patients; premature aging; retarded somatic growth; gonadal dysgenesis; genomic instability manifested as high rates of chromosomal breaks in cultured A-T cells, and clonal translocations involving the immune system genes in patient lymphocytes; profound predisposition to lymphoreticular malignancies (10% of A-T patients develop lymphomas or acute lymphocytic leukemias); and acute sensitivity to the cytotoxic effect of ionizing radiation (IR). The course of A-T is progressive and relentless, and most patients die with respiratory failure or cancer during the second or third decade of life (see Lavin and Shiloh 1997; Rotman and Shiloh 1999, 2000 for recent reviews).

The predominant features of the cellular phenotype of A-T (see above reviews) are genomic instability and a profound defect in responding to DNA damage inflicted by IR and radiomimetic chemicals. The exquisite sensitivity of A-T cells to these agents is underlain by a subtle but clear defect in rejoining of double-strand breaks (DSBs) and failure to activate a variety of other responses normally turned on by this damage. Most notable is defective activation of the cell cycle checkpoints that temporarily arrest cell cycle progression to allow time for repair.

Although defects in responding to DSBs dominate the cellular phenotype of A-T, these cells are characterized by other functional and structural abnormalities as well: premature senescence, cytoskeletal ab-

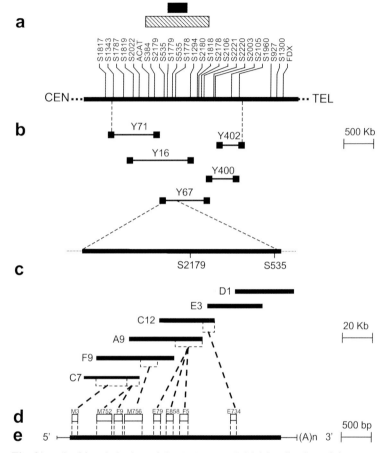

Fig. 2a–e. Positional cloning of the *ATM* gene. Initial localization of the gene on chromosome 11q22–23 (Gatti et al. 1988) led to extensive cloning of the corresponding region in yeast artificial chromosome (**b**) and generation of an extensive map of genetic markers spanning the region (**a**). The *striped* and *solid boxes* denote decreasing intervals for the A-T gene, obtained by repeated linkage analysis carried out by a consortium of several research groups. Subsequent cloning of a critical A-T interval of several hundred kb in cosmid vectors (**c**) allowed the use of gene hunting methods, which yielded cDNA fragments and individual exons (**d**), some of which converged to a single cDNA clone (**e**). This cDNA clone turned out to represent about half of the transcript of a large gene that was found to harbor mutations in A-T patients. (Reproduced with modifications from Savitsky et al. 1995a)

normalities, defects in calcium-dependent signal transduction, defects in signaling cascades activated by the B-cell receptor (observed in A-T lymphocytes), and aberrant potassium currents. Constitutive alterations in chromatin structure were also observed, as well as specific defects in maintaining telomere length, and abnormal associations between chromosome ends and between the telomeres and the nuclear matrix.

This myriad of phenotypic features, all of which are clearly associated with one protein, have made the search for the elusive A-T gene extremely attractive. The search was carried out by several groups over a period of 7 years, beginning with the localization of the A-T gene on chromosome 11q22–23 by Gatti et al. (1988), and the actual isolation of the gene by Savitsky et al. (1995a,b). The positional cloning of the gene, designated *ATM* (A-T, mutated), typically involved all the elements of the classical positional cloning paradigm, with extensive generation of genomic contigs and genetic markers, and application of gene hunting strategies (Savitsky et al. 1995a,b; Fig. 2). Clearly, the current availability of these reagents for the entire human genome has recently made such endeavors considerably simpler.

The *ATM* gene, spanning about 150 kb of genomic DNA, turned out to be quite a highly structured one, with 66 exons constituting a 13 kb mRNA with an open reading frame of about 9.2 kb (Uziel et al. 1996; Platzer et al. 1997). Importantly, the mutations that cause A-T are mainly null alleles, most of which truncate the ATM protein (Gilad et al. 1996; Concannon and Gatti 1997). The discovery of the *ATM* gene led to the generation of a series of animal models of A-T by targeted inactivation of the murine homolog gene, *Atm* (Barlow et al. 1996; Elson et al. 1996; Xu et al. 1996; Herzog et al. 1998). These animals have contributed further information about the functions of *ATM*'s product. All strains of Atm-deficient mice recapitulate most of the features of the human disease, including growth retardation, immunodeficiency, sterility, radiosensitivity, and a striking propensity to thymic lymphomas. Their cerebellar phenotype is, however, barely distinguished compared to the human one.

3 From Gene Back to Phenotype:
Functional Analysis of the ATM Protein

3.1 The Product of the A-T Gene

The *ATM* gene encodes a large protein of 3,056 amino acids with a carboxy-terminal domain of about 350 residues that has high similarity to the catalytic subunit of phosphatidylinositol 3-kinases (PI3-kinases) (Savitsky et al. 1995a,b). These characteristics place ATM within a family of large proteins that were identified in organisms ranging from yeast to mammals, all of which contain the PI3-kinase-like domain, and are involved to various extents in controlling genomic stability, cellular responses to genotoxic stress, and cell cycle progression (reviewed by Rotman and Shiloh 1998, 1999). Notable members of this group in the budding yeast *Saccharomyces cerevisiae* are scTel1p, involved in maintaining telomere length, and scMec1p, which controls several cell cycle checkpoints induced by DNA damage or replication arrest. In the fission yeast *Schizosaccharomyces pombe*, the spRad3p protein has functions similar to those of scMec1p.

Despite the resemblance to lipid kinases, several members of this family, including ATM (see below), were found to possess a serine/threonine protein kinase activity. An extensively characterized member is the catalytic subunit of the DNA-dependent protein kinase (DNA-PKcs), which is activated in vitro by DNA ends, and is involved in processing double-strand breaks via the non-homologous end-joining pathway (Jeggo 1997). However, the roles and substrates of ATM and DNA-PK are clearly separate and different. Another mammalian PI3-kinase-like protein, which is much closer to ATM functionally, is the human homolog of spRad3p, the ATR protein kinase whose responsibilities are partially redundant with those of ATM (Cliby et al. 1998; Tibbetts et al. 1999; reviewed by Shiloh 2001).

ATM is a ubiquitously expressed 370 kDa phosphoprotein (reviewed by Canman and Lim 1998; Lavin and Khanna 1999; Rotman and Shiloh 1999). Cellular localization of ATM is variable; it resides primarily in the nucleus of proliferating cells, with smaller amounts detected in microsomal fractions. A portion of extranuclear ATM is associated with peroxisomes (Watters et al. 1999). A more prominent cytoplasmic fraction was noticed in oocytes (Barlow et al. 1998) and in differentiated cells such as cerebellar neurons (Oka and Takashima 1998; Barlow et al. 2000). The subcellular localization of ATM, as well as the processes defective in A-T cells, clearly point to nuclear as well as cytoplasmic functions of this protein in somatic cells. A possible indication of a cytoplasmic function is the association of ATM with β-adaptin, which is involved in protein sorting and endocytosis (Lim et al. 1998).

ATM may have different roles in different tissues and cell types in the nervous system. Even the degree of its involvement in radiation responses may be cell type-dependent. For example, while various types of neurons in the developing CNS of *Atm –/–* mice are less susceptible to radiation-induced apoptosis than in control animals, Atm-deficient thymocytes are more susceptible to this effect, and astrocytes show no radiation sensitivity but clearly senesce prematurely (Herzog et al. 1998; Gosink et al. 1999). Atm-deficient mice show specific degeneration of dopaminergic nigro-striatal neurons and their terminals in the striatum, with corresponding locomotor abnormalities (Eilam et al. 1998). Atm levels also vary considerably between various parts of the developing and mature CNS (Soares et al. 1999).

ATM plays a role in meiotic recombination in germ-line cells, evidenced by the extensive chromosomal fragmentation followed by apoptotic death at early prophase I that occur in Atm-deficient mice. Indeed, in normal meiotic cells, Atm is associated with synapsed chromosomal axes, where it co-localizes with additional proteins involved in meiotic recombination (reviewed by Lavin and Khanna 1999; Rotman and Shiloh 1999, 2000). These observations underscore the importance of the cellular context for ATM functions.

3.2 A Major Role in Cellular Responses to DNA Double-Strand Breaks

ATM has a critical role in the complex cellular response to the induction of DSBs in the DNA. DSBs are formed in the course of natural processes, such as meiotic recombination or the processing of the immune system genes, and may be induced by endogenous metabolites or exogenous DNA damaging agents. DSBs are handled by a co-ordinated system that repairs the damage, temporarily alters cellular metabolism, and slows the course of cellular life while the damage is repaired. Alterna-

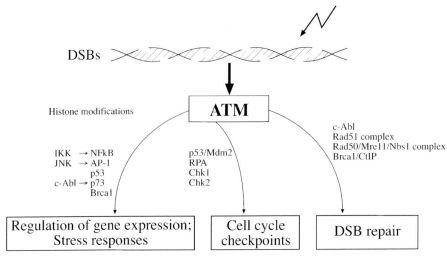

Fig. 3. A summary of cellular responses to DSBs mediated by ATM. Major responses are the repair of DSBs and temporary arrest of the cell cycle via the cell cycle checkpoints. However, numerous other processes are initiated following the induction of DSBs, and involve a profound alteration in the gene expression profile of the cell. Most of these alterations are dependent on ATM. See text for details. (Reproduced from Shiloh 2000 with permission of Cold Spring Harbor Laboratory Press)

tively, the cell may succumb to an overload of damage and opt for apoptosis via other signaling pathways (Dasika et al. 1999; Karran 2000; Lowndes and Murguia 2000). Two major repair systems for DSBs, homologous recombination (HR) and non-homologous end joining (NHEJ), are responsible in eukaryotic organisms for the restoration of DNA continuity at the DSB site, and each one of them is composed of multi-protein complexes (reviewed by Karran 2000). Major players in the activation of the cell cycle checkpoints are the p53 protein, which has a predominant role on the G1/S checkpoint and also takes part in the G2/M checkpoint (reviewed by Oren 1999), replication protein A (RPA), thought to be involved in the S-phase checkpoint (Iftode et al. 1999), and the Chk1 and Chk2 protein kinases that play a central role in the G2/M checkpoint, while Chk2 is also involved in the G1/S checkpoint (Matsuoka et al. 1998, 2000; Chatuvredi et al. 1999; Chehab et al. 1999, 2000; Hirao et al. 2000; Shieh et al. 2000).

Certain proteins seem to function in more than one mechanism. For example, the Brca1 protein, the product of one of the breast cancer susceptibility genes, appears to take part in a variety of damage-induced functions ranging from transcription-coupled repair of oxidative damage and DSB repair to up-regulation of damage-induced genes such as *TP21* and *GADD45* (reviewed by Welcsh et al. 2000). In addition, major transcription factors involved in stress responses, such as AP-1 and NFκB, also respond to DNA damage, and in the case of DSBs their response is mediated by ATM (see below). ATM is clearly a major co-ordinator of this intricate system in the early phase following damage infliction, reacting immediately and swiftly by mounting a concerted action on these pathways (Fig. 3).

3.3 Understanding ATM's Mode of Function by Identifying Its Substrates

The catalytic activity of ATM, an important key to understanding its functions, was recently shown to be that of a serine/threonine protein kinase (Banin et al. 1998; Canman et al. 1998; Khanna et al. 1998). This activity is enhanced several-fold immediately following treatment of cells with DNA breaking agents (Banin et al. 1998; Canman et al. 1998) by a yet unknown mechanism. Identification of the in vivo targets of this activity is crucial to identifying specific signaling pathways downstream of ATM.

Several key players in the salvage system induced by DSBs (Fig. 3) are immediately phosphorylated in an ATM-dependent manner. Interestingly, in order to ensure the co-ordinated response of several pathways, ATM acts simultaneously on several key targets, sometimes more than one in the same pathway, and on proteins that sit at the junctions of several pathways (Fig. 4). Thus, ATM directly phosphorylates p53 on Ser15 (Banin et al. 1998; Canman et al. 1998; Khanna et al. 1998), thereby enhancing its transactivation capacity (Dumaz and Meek 1999), and activates a process that ends in dephosphorylation of Ser376 of p53 (Waterman et al. 1998). At the same time, ATM phosphorylates p53's inhibitor, Mdm2, on Ser395 (Khosravi et al. 1999; Maya et al., submitted). Mdm2 binds to p53, inhibits its transactivation activity and exports it from the nucleus to the cytoplasm while serving as the E3 ligase in p53's degradation via the

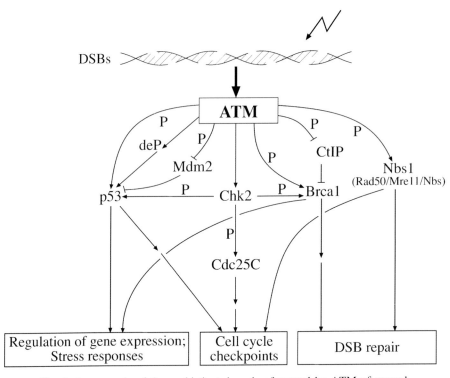

Fig. 4. An example of the sophisticated mode of control by ATM of several pathways induced by DSBs. ATM is responsible, directly or indirectly, for inducing post-translational modifications in key proteins in the network of cellular responses to DSBs. Sometimes, several ATM-dependent modifications are induced in the same protein. See text for details. (Reproduced from Shiloh 2000 with permission of Cold Spring Harbor Laboratory Press)

proteasome pathway (reviewed by Oren 1999; Momand et al. 2000). Although the phosphorylation of Mdm2 by ATM on Ser395 does not inhibit p53-Mdm2 interaction, it does interfere with Mdm2's ability to export p53 from the nucleus (Maya et al., submitted). In parallel, ATM phosphorylates and activates the Chk2 kinase (Chaturvedi et al. 1999; Matsuoka et al. 1998, 2000; Brown et al. 1999), which in turn phosphorylates p53 on Ser20, thereby interrupting its interaction with Mdm2 (Chehab et al. 1999, 2000; Hirao et al. 2000; Shieh et al. 2000).

An analogous mechanism operates in another arm of this network where the Brca1 is activated to function in DSB repair and in the activation of genes involved in the survival response. Several phosphorylation events activate Brca1 at the same time: ATM phosphorylates it directly on several sites (Cortez et al. 1999; Gatei et al. 2000a); ATM-activated Chk2 phosphorylates it at another site (Lee at el. 2000); and ATM-mediated phosphorylation of CtIP, an inhibitor of Brca1's transactivating ability, interferes with its interaction with Brca1 (Li et al. 2000) (Fig. 2). These elegant and effective mechanisms for controlling functional networks may be applicable to other ATM-mediated functions.

Another dual-function protein is Nbs1, which is part of the DSB repair complex Rad50/Mre11/Nbs1 and is also involved in the activation of the S-phase checkpoint following ionizing radiation damage (Petrini 2000; Fig. 3). Nbs1 is phosphorylated by ATM on several sites, and this phosphorylation is essential for both its roles in the DSB response pathways (Gatei et al. 2000b; Lim et al. 2000; Wu et al. 2000; Zhao et al. 2000). The elucidation of the functional link between ATM and the Rad50/Mre11/Nbs1 complex also explained the relationship between A-T and two other genetic disorders phenotypically related to A-T. The first is the Nijmegen breakage syndrome (NBS), which is characterized by immunodeficiency, genomic instability, radiation sensitivity, and cancer predisposition (Digweed et al. 1999). The responsible gene encodes the Nbs1 member of this complex. The second disorder, "ataxia-telangiectasia-like disorder" (ATLD), is a moderate version of classical A-T and is caused by mutations in the gene for the Mre11 protein (Stewart et al. 1999). Clearly, the classical A-T phenotype spans the phenotypes of both disorders plus additional features unique to A-T, and this hierarchy correlates with the position of ATM in the DSB signaling cascades, above the Rad50/Mre11/Nbs1 complex.

A notable downstream effector of ATM is c-Abl, a nuclear tyrosine kinase implicated in the survival and apoptotic responses to various stresses including DSBs (Shaul 2000). ATM and c-Abl interact physically (Shafman et al. 1997), and following ionizing radiation c-Abl is activated via its phosphorylation in an ATM-dependent manner (Baskaran et al. 1997). A distinct apoptotic pathway mediated by c-Abl involves the p73 protein, a member of the p53-like family (Lohrum and Vousden 2000). Following ionizing radiation treatment, p73 is phosphorylated in a c-Abl- and ATM-dependent manner, while after cisplatin treatment its level rises in dependence on c-Abl and the mismatch repair protein MLH1 (Agami et al. 1999; Gong et al. 1999; Yuan et al. 1999). The involvement of ATM in the latter pathway is not clear. Another pathway mediated by c-Abl is the assembly of the homologous recombination repair complex containing the Rad51 and Rad52 proteins, which follows radiation-induced phosphorylation of Rad51 in an ATM- and c-Abl-dependent manner (Chen et al. 1999). These observations clearly point to the c-Abl protein as a downstream mediator of several ATM-dependent pathways, but the exact mode of transmission of the signal from ATM to c-Abl remains to be delineated.

3.4 ATM and Gene Expression

While many of the processes initiated by DNA damage are initially represented by signal transmission via post-translational modifications of proteins, a considerable part of them end up in alterations in gene expression. The change in the cellular transcriptome following DNA damage is impressive, and includes hundreds of genes involved in all aspects of cellular physiology (Jelinsky and Samson 1999; Rashi-Elkeles et al., submitted; R. Rolig, personal communication). A close look at the list of proteins involved in ATM-mediated pathways (Fig. 3) reveals several major transcription factors, such as p53, AP-1, and NFκB. AP-1 and NFκB are best known for their activation by mitogenic stimuli or various stress agents, where the primary signals are initiated at the cellular membrane or in the cytoplasm. However, these transcription factors are also activated by DNA damaging agents, including those that induce DSBs, in which case the damage to DNA is the primary source of the signal coming from the nucleus.

The c-Jun proteins, which together with the c-fos proteins comprise the AP-1 family of transcription factors, can be activated via their phosphorylation by the c-Jun N-terminal kinases (JNKs), which are a stress-responsive arm of the mitogen-activated protein (MAP) kinase system (Tibbles and Woodgett 1999). The activation of these pathways by DSBs is ATM dependent, since IR-induced JNK activation is defective in A-T cells (Shafman et al. 1995; Lee et al. 1998a). Notably, JNK activation by mitogenic stimuli is normal in the absence of ATM. The transcription factor NF-κB is activated by a broad array of stress-related signals, including inflammatory cytokines and DNA damage (Mercurio and Manning 1999). In most cells, NF-κB is sequestered in the cytoplasm in an inactive form due to its tight association with its inhibitor, IκB. Activation of NF-κB is achieved through signal-induced phosphorylation of IκB at specific N-terminal serine residues by the multi-protein complex IκB-kinase (IKK). This phosphorylation triggers IκB degradation via the ubiquitin-proteasome pathway, resulting in NF-κB translocation into the nucleus (Karin 1999). ATM appears to mediate the response of NF-κB to DSBs, since its activation by IR and the topoisomerase inhibitor camptothecin is reduced or abolished in A-T cells (Lee et al. 1998b; Piret et al. 1999). Recently, Li et al. (2001) found that functional ATM is required for the activation of IKK, and consequently NF-κB, by ionizing radiation and radiomimetic chemicals. However, the absence of ATM did not affect the induction of this pathway in response to pro-inflammatory stimuli, such as tumor necrosis factor (TNF), phorbol myristate acetate (PMA) and lipopolysaccharide (LPS). c-Abl and DNA-PK were not required for the induction of NF-κB by IR. These few examples underscore the potential ability of ATM to modulate the expression of numerous genes by affecting the activity of major stress-responsive transcription factors.

4 Conclusions

The quest for the A-T gene and the current phase of ATM research demonstrate at the single-gene level the typical move from "sequence genomics" to functional genomics. Clearly, understanding the function of a single gene (protein) will entail, in many cases, understanding the functions of many others, which are functionally related to it. ATM may be a particularly salient example of this phenomenon, but the principle will hold, to varying degrees, for many proteins. Complete functional understanding of a single gene at the cellular and organismal levels will thus require knowledge of numerous components of a web of physical and functional interactions that together shape our biology.

Acknowledgements. The author is indebted to the members of the A-T group at Tel Aviv University for their dedication to A-T research, and to numerous collaborators and contributors of reagents and expertise to this research. Work in the author's laboratory was supported by the A-T Medical Research Foundation, the A-T Children's Project, the Thomas Appeal (A-T Medical Research Trust), the A-T Society, and the National Institutes of Health.

References

Agami R, Blandino G, Oren M, Shaul Y (1999) Interaction of c-Abl and p73α and their collaboration to induce apoptosis. Nature 399:809–813

Banin S, Moyal L, Shieh S-Y, Taya Y, Anderson CW, Chessa L, Smorodinsky NI, Prives C, Reiss Y, Shiloh Y, Ziv Y (1998) Enhanced phosphorylation of p53 by ATM in response to DNA damage. Science 281:1674–1677

Barlow C, Liyanage M, Moens PB, Tarsouna M, Nagashima K, Brown K, Rottinghaus S, Jackson SP, Tagle D, Ried, T, Wynshaw-Boris A (1998) Atm deficiency results in severe meiotic disruption as early as leptonema of prophase I. Development 125:4007–4017

Barlow C, Dennery PA, Shigenman MK, Smith MA, Morrow JD, Roberts LJ, Wynshaw-Boris A, Levine RL (1999) Loss of the ataxia-telangiectasia gene product causes oxidative damage in target tissues. Proc Natl Acad Sci USA 96:9915–9919

Barlow C, Ribaut-Barassin C, Zwingman C, Pope AJ, Brown KD, Larson D, Harrington EA, Haeberle AM, Mariani J, Eckhouse M, Herrup K, Baili Y, Wynshaw-Boris A (2000) ATM is a cytoplasmic protein in mouse brain required to prevent lysosomal accumulation. Proc Natl Acad Sci USA 97:871–876

Baskaran R, Wood LD, Whitaker LL, Canman CE, Morgan SE, Xu Y, Barlow C, Baltimore D, Wynshaw-Boris A, Kastan MB, Wang JYJ (1997) Ataxia telangiectasia mutant protein activates c-Abl tyrosine kinase in response to ionizing radiation. Nature 387:516–519

Brown AL, Lee C-H, Schwarz JK, Mitiku N, Piwnica-Worms H, Chung JH (1999) A human Cds1-related kinase that functions downstream of ATM protein in the cellular response to DNA damage. Proc Natl Acad Sci USA 96:3745–3750

Canman CE, Lim D-S (1998) The role of ATM in DNA damage and cancer. Oncogene 17:3301–3308

Canman CE, Lim D-S, Cimprich KA, Taya Y, Tamai K, Sakaguchi K, Appella E, Kastan MB, Siliciano JD (1998) Activation of the ATM kinase by ionizing radiation and phosphorylation of p53. Science 281:1677–1679

Chaturvedi P, Eng WK, Zhu Y, Mattern MR, Mishra R, Hurle MR, Zhang X, Annan RS, Lu Q, Faucette LF, Scott GF, Li X, Carr SA, Johnson RK, Winkler JD, Zhou B-BS (1999) Mammalian Chk2 is downstream effector of the ATM-dependent DNA damage checkpoint pathway. Oncogene 18:4047–4054

Chehab NH, Malikzay A, Stavridi ES, Halazonetis TD (1999) Phosphorylation of Ser-20 mediates stabilization of human p53 in response to DNA damage. Proc Natl Acad Sci USA 96:13777–13782

Chehab NH, Malikzay A, Appel M, Halazonetis TD (2000) Chk2/hCds1 functions as a DNA damage checkpoint in G_1 by stabilizing p53. Genes Dev 14:278–288

Chen G, Yuan S-SF, Liu W, Xu Y, Trujillo K, Song B, Cong F, Goff SP, Wu Y, Arlinghaus R, Baltimore D, Gasser PJ, Park MS, Sung P, Lee EY-HP (1999) Radiation-induced assembly of Rad51 and Rad52 recombination complex requires ATM and c-Abl. J Biol Chem 274:12748–12752

Cliby WA, Roberts CJ, Cimprich KA, Stringer CM, Lamb JR, Schreiber SL, Friend SH (1998) Overexpression of a kinase-inactive ATR protein causes sensitivity to DNA-damageing agents and defects in cell cycle checkpoints. EMBO J 17:159–169

Concannon P, Gatti RA (1997) Diversity of *ATM* mutations detected in patients with ataxia-telangiectasia. Hum Mutat 10:100–107

Cortez D, Wang Y, Qin J, Elledge SJ (1999) Requirement of ATM-dependent phosphorylation of Brca1 in the DNA damage response to double-strand breaks. Science 286:1162–1166

Dasika GK, Lin S-CJ, Zhao S, Sung P, Tomkinson A, Lee EY-HP (1999) DNA damage-induced cell cycle checkpoints and DNA strand break repair in development and tumorigenesis. Oncogene 18:7883–7899

Digweed M, Reis A, Sperling K (1999) Nijmegen breakage syndrome: consequences of defective DNA double strand break repair. BioEssays 21:649–656

Dumaz N, Meek DW (1999) Serine15 phosphorylation stimulates p53 transactivation but does not directly influence interaction with HDM 2. EMBO J 18:7002–7010

Eilam R, Peter Y, Elson A, Rotman G, Shiloh Y, Groner Y, Segal M (1998) Selective loss of dopaminergic nigro-striatal neurons in brains of Atm-deficient mice. Proc Natl Acad Sci USA 95:12653–12656

Elson A, Wang Y, Daugherty CJ, Morton CC, Zhou F, Campos-Torres J, Leder P (1996) Pleiotropic defects in ataxia-telangiectasia protein- deficient mice. Proc Natl Acad Sci USA 93:13084–13089

Gatei M, Scott SP, Fillippovitch I, Soronika N, Lavin MF, Weber B, Khanna KK (2000a) Role for ATM in DNA damage-induced phosphorylation of BRCA1. Cancer Res 60:3299–3304

Gatei M, Young D, Cerosaletti KM, Desai-Mehta A, Spring K, Kozlov S, Lavin MF, Gatti RA, Concannon P, Khanna KK (2000b) ATM-dependent phosphorylation of nibrin in response to radiation exposure. Nature Genet 25:115–119

Gatti RA, Berkel I, Boder E, Braedt G, Charmley P, Concannon P, Ersoy F, Foroud T, Jaspers NG, Lange K et al (1988) Localization of an ataxia-telangiectasia gene to chromosome 11q22–23. Nature 336:577–80

Gilad S, Khosravi R, Uziel T, Ziv Y, Rotman G, Savitsky K, Smith S, Chessa L, Harnik R, Shkedi D, Frydman M, Sanal O, Portnoi S, Goldwicz Z, Jaspers NGJ, Gatti RA, Lenoir G, Lavin MF, Tatsumi K, Wegner RD, Shiloh Y, Bar-Shira A (1996) Predominance of null mutations in ataxia-telangiectasia. Hum Mol Genet 5:433–439

Gong J, Costanzo A, Yang H-Q, Melino G, Kaelin WG Jr, Levrero M, Wang JYJ (1999) The tyrosine kinase c-Abl regulates p73 in apoptotic response to cisplatin-induced DNA damage. Nature 399:806–809

Gosink EC, Chong M, McKinnon PJ (1999) Ataxia telangiectasia mutated deficiency affects astrocyte growth but not radiosensitivity. Cancer Res 59:5294–5298

Herzog K-H, Chong MJ, Kapsetaki M, Morgan JI, McKinnon P (1998) Requirement for Atm in ionizing radiation-induced cell death in the developing central nervous system. Science 280:1089–1091

Hirao A, Kong Y-Y, Matsuoka S, Wakeham A, Ruland J, Yoshida H, Liu D, Elledge SJ, Mak TW (2000) DNA damage-induced activation of p53 by the checkpoint kinase Chk2. Science 287:1824–1827

Iftode C, Daniely Y, Borowiec JA (1999) Replication protein A (RPA): the eukaryotic SBB. Crit Rev Biochem Mol Biol 34:141–180

Jeggo PA (1997) DNA-PK: at the cross-roads of biochemistry and genetics. Mutat Res 384:1–14

Khanna KK, Keating KE, Kozlov S, Scott S, Gatei M, Hobson K, Taya Y, Gabrielli B, Chan D, Lees-Miller SP, Lavin MF (1998) ATM associates with and phosphorylates p53: mapping the region of interaction. Nature Genet 20:398–400

Jelinsky SA, Samson LD (1999) Global response of *Saccharomyces cerevisiae* to an alkylating agent. Proc Natl Acad Sci USA 96:1486–1491

Karin M (1999) How NF-κB is activated: the role of the IκB kinase (IKK) complex. Oncogene 18:6867–6874

Karran P (2000) DNA double strand break repair in mammalian cells. Curr Opin Genet Dev 10:144–150

Khosravi R, Maya R, Gottlieb T, Oren M, Shiloh Y, Shkedy D (1999) Rapid ATM-dependent phosphorylation of MDM 2 precedes p53 accumulation in response to DNA. Proc Natl Acad Sci USA 96:14973–14977

Lavin MF, Shiloh Y (1997) The genetic defect in ataxia-telangiectasia. Annu Rev Immunol 15:177–202

Lavin MF, Khanna KK (1999) ATM: the protein encoded by the gene mutated in the radiosensitive syndrome ataxia-telangiectasia. Int J Radiat Biol 75:1201–1214

Lee SA, Dristschilo A, Jung M (1998a) Impaired ionizing radiation- induced activation of a nuclear signal essential for phosphorylation of c-Jun by dually phosphorylated c-Jun amino-terminal kinases in ataxia telangiectasia fibroblasts. J Biol Chem 273:32889–32894

Lee S-J, Dimtchev A, Lavin MF, Dristschilo A, Jung M (1998b) A novel ionizing radiation-induced signaling pathway that activates the transcription factor NF-κB. Oncogene 17:1821–1826

Lee J-S, Collins KM, Brown AL, Lee C-H, Chung JH (2000) HCds1-mediated phosphorylation of BRCA1 regulates the DNA damage response. Nature 404:201–204

Li S, Ting NSY, Zheng L, Chen P-L, Ziv Y, Shiloh Y, Lee EY-H, Lee W-H (2000) Functional link of Ataxia-telangiectasia gene product and BRCA1 in DNA damage response. Nature 406:210–215

Li N, Banin S, Ouyang H, Li GC, Courtois G, Shiloh Y, Karin M, Rotman G (2001) ATM is required for IKK activation in response to DNA double strand breaks. J Biol Chem 276:8898–8903

Lim D-S, Kim S-T, Xu B, Maser RS, Lin J, Petrini JHJ, Kastan MB (2000) ATM phosphorylates p95/nbs1 in an S-phase checkpoint pathway. Nature 404:613–617

Lim D-S, Kirsch D, Canman CE, Ahn J-H, Ziv Y, Newman LS, Darnell R, Shiloh Y, Kastan MB (1998) ATM binds to β-adaptin in cytoplasmic vesicles. Proc Natl Acad Sci USA 95:10146–10151

Lohrum MAE, Vousden KH (2000) Regulation and function of the p53-related proteins: same family, different rules. Trends Cell Biol 10:197–202

Lowndes NF, Murguia JR (2000) Sensing and responding to DNA damage. Curr Opin Genet Dev 10:17–25

Matsuoka S, Huang M, Elledge SJ (1998) Linkage of ATM to cell cycle regulation by the Chk2 protein kinase. Science 282:1893–1897

Matsuoka S, Rotman G, Ogawa A, Shiloh Y, Tamai K, Elledge S (2000) ATM phosphorylates Chk2 in vivo and in vitro. Proc Natl Acad Sci USA 97:10389–10394

Mercurio F, Manning AM (1999) NF-κB as a primary regulator of the stress response. Oncogene 18:6163–6171

Momand J, Wu HH, Dasgupta G (2000) MDM 2 – master regulator of the p53 tumor suppressor protein. Gene 242:15–29

Oka A, Takashima S (1998) Expression of the ataxia-telangiectasia gene product in human cerebellar neurons during development. Neurosci Lett 252:195–198

Oren M (1999) Regulation of p53 tumor suppressor protein. J Biol Chem 274:36031–36034

Petrini JHJ (2000) The Mre11 complex and ATM: collaborating to navigate S phase. Curr Opin Cell Biol 12:293–294

Piret B, Schoonbroodt S, Piette J (1999) The ATM protein is required for sustained activation of NF-κB following DNA damage. Oncogene 18:2261–2271

Platzer M, Rotman G, Bauer D, Uziel T, Savitsky K, Bar-Shira,A, Gilad S, Shiloh Y, Rosenthal A (1997) Ataxia-telangiectasia locus: analysis of 184 kb of genomic DNA containing the entire ATM gene. Genome Res 7:592–605.

Rotman G, Shiloh Y (1998) ATM: from gene to function. Hum Mol Genet 7:1555–1563

Rotman G, Shiloh Y (1999) ATM: a mediator of multiple responses to genotoxic stress. Oncogene 18:6135–6144

Rotman G, Shiloh Y (2000) ATM: at the crossroads of DNA damage response, cell cycle control, genome stability and cancer. In: Erlich M (ed) DNA alterations in cancer: genetic and epigenetic changes. Eaton Publishing, Natick, MA, pp 227–240

Savitsky K, Bar-Shira A, Gilad S, Rotman G, Ziv Y, Vanagaite L, Tagle DA, Smith S, Uziel T, Sfez S, Ashkenazi M, Pecker I, Frydman M, Harnik R, Patanjali SR, Simmons A, Clines GA, Sartiel A, Gatti RA, Chessa L, Sanal O, Lavin MF, Jaspers NGJ, Taylor AMR, Arlett CF, Miki T, Weissman S, Lovett M, Collins FS, Shiloh Y (1995a) A single ataxia telangiectasia gene with a product similar to PI-3 kinase. Science 268:1749–1753

Savitsky K, Sfez S, Tagle D, Ziv Y, Sartiel A, Collins FS, Shiloh Y, Rotman G (1995b) The complete sequence of the coding region of the ATM gene reveals similarity to cell cycle regulators in different species. Hum Mol Genet 4:2025–2032

Shafman TD, Saleem A, Kyriakis J, Weichselbaum R, Kharbanda S, Kufe DW (1995) Defective induction of stress-activated protein kinase activity in ataxia-telangiectasia cells exposed to ionizing radiation. Cancer Res 55:3242–3245

Shafman T, Khanna KK, Kedar P, Spring K, Kozlov S, Yen T, Hobson K, Gatei M, Zhang N, Watters D, Egerton M, Shiloh Y, Kharbanda S, Kufe D, Lavin MF (1997) Interaction between ATM protein and c-Abl in response to DNA damage. Nature 387:520–523

Shaul Y (2000) c-Abl: activation and nuclear targets. Cell Death Differ 7:10–16

Shieh S-Y, Ahn J, Tamai K, Taya Y, Prives C (2000) The human homologs of checkpoint kinases Chk1 and Cds1 (Chk2) phosphorylate p53 at multiple DNA damage-inducible sites. Genes Dev 14:289–300

Shiloh Y (2000) ATM: Sounding the double-strand break alarm. Cold Spring Harbor Symposia on Quantitative Biology, "Biological responses to DNA damage". 65:527–533

Shiloh Y (2001) ATM and ATR: networking cellular responses to DNA damage. Curr Opin Genet Dev (in press)

Soares HD, Morgan JI, McKinnon PJ (1999) *Atm* expression patterns suggest a contribution from the peripheral nervous system to the phenotype of ataxia-telangiectasia. Neuroscience 86:1045–1054

Stewart GS, Maser RS, Stankovic T, Bressan DA, Kaplan MI, Jaspers NG, Raams A, Byrd PJ, Petrini JH, Taylor AM (1999) The DNA double-strand break repair gene hMRE11 is mutated in individuals with an ataxia-telangiectasia-like disorder. Cell 10:577–587

Tibbetts RS, Brumbaugh KM, Williams JM, Cliby WA, Shieh S-Y, Taya Y, Prives C, Abraham RT (1999) A role for ATR in the DNA damage-induced phosphorylation of p53. Genes Dev 13:152–157

Tibbles LA, Woodgett JR (1999) The stress-activated protein kinase pathways. Cell Mol Life Sci 55:1230–1254

Uziel T, Savitsky K, Platzer M, Ziv Y, Helbitz T, Nehls M, Boehm T, Rosenthal A, Shiloh Y, Rotman G (1996) Genomic organization of the ATM gene. Genomics 33:317–320

Waterman MJF, Stavridi ES, Waterman JLF, Halazonetis TD (1998) ATM-dependent activation of p53 involves dephosphorylation and association with 14-3-3 proteins. Nature Genet 19:175–178

Watters D, Kedar P, Spring K, Chen P, Gatei M, Birrell G, Garrone B, Srinivas P, Crane DI, Lavin MF (1999) Localization of a portion of extranuclear ATM to peoxisomes. J Biol Chem 274:34277–34282

Welcsh PL, Owens KN, King M-C (2000) Insights into the functions of BRCA1 and BRCA2. TIG 16:69–74

Wu X, Ranganathan V, Weisman DS, Heine WF, Ciccone DN, O'Neill TB, Crick KE, Pierce KA, Lane WS, Tahbun G, Livingston DM, Weaver DT (2000) ATM phosphorylation of Nijmegen breakage syndrome protein is required in a DNA damage response. Nature 405:477–482

Xu Y, Ashley T, Brainerd EE, Bronson RT, Meyn SM, Baltimore D (1996) Targeted disruption of ATM leads to growth retardation, chromosomal fragmentation during meiosis, immune defects and thymic lymphoma. Genes Dev 10:2411–2422

Yuan Z-M, Shioya H, Ishiko T, Sun X, Gu J, Huang Y, Lu H, Kharbanda S, Weichselbaum R, Kufe D (1999) p73 is regulated by tyrosine kinase c-Abl in the apoptotic response to DNA damage. Nature 399:814–817

Zhao S, Weng Y-C, Yuan S-SF, Lin Y-T, Hsu H-C, Lin S-CJ, Gerbino E, Song M-h, Zdzienicka MZ, Gatti RA, Shay JW, Ziv Y, Shiloh Y, Lee EY-HP

(2000) Functional link between ataxia-telangiectasia and Nijmegen break-age syndrome gene products. Nature 405:473–477

5 The Genetics of Deafness: A Model for Genomic and Biological Complexity

K.B. Avraham

1 Introduction

Advanced genomic technologies have led to the isolation of many genes involved in human disease, and the study of the human genome is now moving towards understanding the function of the proteins these genes encode (functional genomics). The intricate structure and multiple cell types of the inner ear require a range of proteins with different functions, including maintenance of structural integrity, neuronal innervation, and mechanoelectrical transduction. There has been remarkable progress in the field of hereditary hearing loss over the past 5 years in elucidating the molecular basis of hearing loss. This has been no easy task, as human deafness is extremely heterogeneous. Not only is there great variability in the clinical features of human hearing loss (HL), but mutations in the same genes can contribute to syndromic, nonsyndromic, prelingual, and progressive deafness. This variability complicates genotype–phenotype correlations and hence our understanding of how mutations lead to inner ear pathogenesis.

This chapter is based loosely on a talk presented in Berlin on 1–2 November 2000 at the workshop entitled, "The Human Genome: Biology and Medicine." The aim is to highlight areas in which research is being done in my own laboratory, putting the results in context of the field in general (the unconventional myosin VI, the *DFNA22* human deafness locus and Snell's waltzer mouse, the POU4F3 transcription factor and the *DFNA15* human deafness locus, and inherited connexin 26 mutations associated with human non-syndromic hearing loss). I will summarize the genes identified in humans, the techniques used to do so, and the advantages offered by the mouse as a model for human deafness, highlighting several examples of each. This is not meant to be a comprehensive review. Indeed, progress in the past few years has been so tremendous and rapid, listing all genes identified would be a formidable task.

2 Genetic Hearing Loss

2.1 Syndromic Hearing Loss

Over half of all deafness is due to genetic causes and occurs both in association with other symptoms in the form of syndromic hearing loss (SHL), or as an isolated finding, non-syndromic hearing loss (NSHL). There are over 400 clinically defined syndromes that include hearing loss (Online Mendelian Inheritance in Man [OMIM]; see Table 1). In some cases, the hearing loss is a minor feature (e.g. Charcot-Marie-Tooth disease, osteogenesis imperfecta); in others, hearing loss is a major feature. These include Pendred syndrome (with goitre), Waardenburg syndrome (with pigmentary anomalies and widely-spaced eyes), Alport syndrome (with kidney defects), branchio-oto-renal (BOR) syndrome (with craniofacial and kidney defects) and Usher syndrome (with retinitis pigmentosa). The genes for some of the more prevalent forms of SHL have been identified, including pendrin (Pendred syndrome),

Table 1. General Web sites

Web site	Address
Online Mendelian Inheritance in Man (OMIM)	http://www.ncbi.nlm.nih.gov/Omim/
GSF ENU-Mouse Muta-genesis Screen Project	http://www.gsf.de/ieg/groups/enu/mutants/index.html
UK Mouse Genome Centre ENU Muta-genesis Programme	http://www.mgu.har.mrc.ac.uk/mutabase/

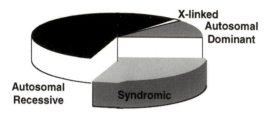

Fig. 1. Genetic heterogeneity of deafness. Approximately 60% of hearing loss is genetic. The *pie* shows proportion of genetic hearing loss that is syndromic and nonsyndromic (and modes of inheritance)

myosin VIIA (Usher syndrome type IB), PAX3 (Waardenburg syndrome type I and III), EYA1 (BOR syndrome) and COL4A5 (X-linked Alport syndrome). It is estimated that 30% of genetic hearing loss is associated with syndromes (Fig. 1).

2.2 Non-syndromic Hearing Loss

Approximately 70% of genetic hearing loss is non-syndromic in nature (Fig. 1). The largest proportion is inherited in an autosomal recessive mode (~80%), ~18% is inherited in an autosomal dominant mode, and ~2% is X-linked. Mitochondrial/maternal inheritance also contributes to a small (1%) proportion of NSHL. Over 100 genes may be involved in

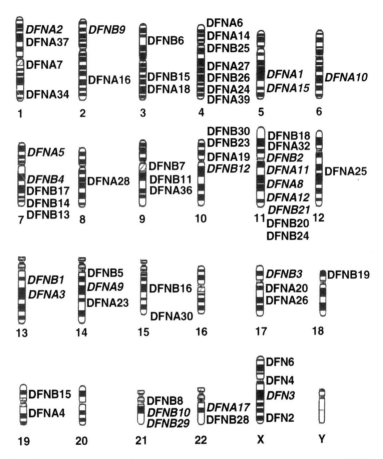

Fig. 2. Loci for non-syndromic hearing loss in the human genome. *DFNA*, autosomal dominant; *DFNB*, autosomal recessive; *DFN*, X-linked. The loci in *grey* have been cloned

Table 2. Inner ear and hearing loss Web sites

Web site	Address
Hereditary Hearing Loss Homepage	http://www.uia.ac.be/dnalab/hhh/
Connexins and Deafness Homepage	http://www.iro.es/deafness/
Table of gene expression in the developing ear	http://www.ihr.mrc.ac.uk/hereditary/ genetable/index.shtml
Inner ear gene expression database	http://www.mgh.harvard.edu/depts/ coreylab/genomics.html
Human Cochlear cDNA library and EST database	http://hearing.bwh.harvard.edu/ cochlearcdnalibrary.htm
Hereditary hearing impairment in mice (HHIM)	http://www.jax.org/research/hhim/
National Institute on Deafness and Other Communication Disorders	www.nidcd.nih.gov

NSHL, and the chromosomal location for 70 have already been found (Fig. 2; updated regularly in the Hereditary Hearing Loss Homepage, see Table 2). This has been possible due to the collection of many families with inherited hearing loss from around the world and the ease of performing genome scans to identify the causative loci. But it remains clear that there are many more genes involved in hearing loss, and another extremely useful way to identify these has been through the use of mouse models.

3 The Mouse as a Model for Human Deafness

At least five genes responsible for human HL were initially identified in the mouse (reviewed in Probst and Camper 1999). An additional ten genes have been implicated in hearing loss in the mouse, and while mutations in these genes have not yet been found in humans, the proteins they encode clearly perform essential roles in the inner ear. Their contribution to our understanding of the auditory system cannot be underestimated. The human and mouse auditory systems are remarkably similar. Furthermore, manipulative genetic tools in the mouse, such as transgenic technologies and mutagenesis, allow scientists to introduce or remove genetic mutations at will, and genes for deafness are no exception (Tymms and Kola 2000). A variety of mouse mutants have been used to identify genes involved in inner ear function, including spontaneously-derived mice, N-ethyl-N-nitrosourea (ENU)-derived mice, and mutants obtained by gene-targeted mutagenesis. Spontaneous mutants arise without any intervention, due to natural causes. Some of these may affect the auditory and vestibular system, and over the years many scientists have taken care to identify and characterize these mutants. Mouse mutations can be induced by radiation or by treating mice with chemicals, such as ENU. Mice are then screened for a specific phenotype, such as deafness or vestibular malfunction, and the genes can then be mapped and identified using different cloning approaches. Gene targeting, or "knockouts," enable us to remove or mutate a specific gene in the mouse and thus study the effects of the gene due to its loss of function. Deafness and vestibular knockout mice are a resource to study specific auditory gene function and morphological characterization of the inner ear.

While the genomic similarities between humans and mice are extremely valuable for the identification of deafness genes, mice also allow us to understand the function of these genes in a way not possible in humans. Due to similarities between humans and mice, phenotypic studies of hearing loss in mice can teach us a great deal about the human auditory and vestibular systems. Techniques such as scanning (SEM) and transmission (TEM) electron microscopy, in situ hybridization, immunochemical studies, electrophysiology and much more, allow us to study the different functions of proteins in the mouse auditory and vestibular systems.

4 Unconventional Myosins in Deafness

Three unconventional myosins are now known to play a role in hearing due to their mutations in mouse mutations (reviewed in Friedman et al. 1999). Myosins are a superfamily of motor proteins defined by their ability to bind actin and hydrolyze adenosine triphosphate (ATP). Unconventional myosins have been implicated in many crucial cellular functions, including endocytosis, ion channel regulation, anchoring of

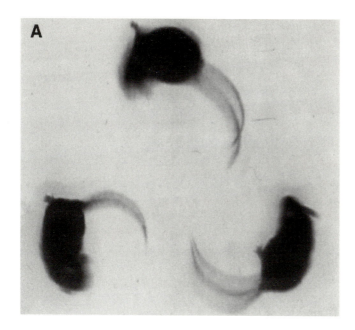

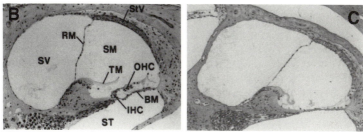

Fig. 3. A A circling Snell's waltzer mouse. **B** Histological section through cochlea of wild-type 6-week-old mouse (Avraham et al. 1995). *SV*, scala vestibuli; *RM*, Reissner's membrane; *SM*, scala media; *TM*, tectorial membrane; *OHC*, outer hair cells; *BM*, basilar membrane; *IHC*, inner hair cells; *ST*, scala tympani, stria vascularis. **C** Neuroepithelial degeneration in the organ of Corti of *sv/sv* cochleas

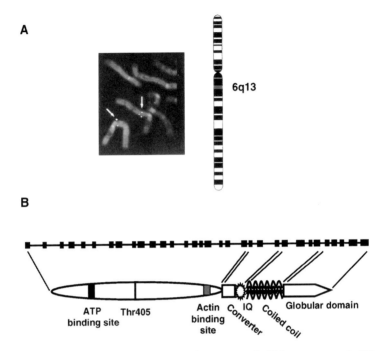

Fig. 4. A Human myosin VI maps to chromosome 6q13 (*arrows*, myosin VI) (Avraham et al. 1997). **B** Schematic diagram of the human myosin VI protein (Ahituv et al. 2000)

stereocilia and vesicle movement (Mermall et al. 1998). In all of these cases the identification of the mouse gene was instrumental in the identification of the human deafness gene.

The shaker1 (*sh1*) mouse is a spontaneously-induced mutant exhibiting deafness, hyperactivity, head-tossing and circling. Usher syndrome type 1B (*USH1B*) was mapped to the homologous chromosomal region in mice. Once the *sh1* gene was discovered by a positional cloning approach (Gibson et al. 1995), it quickly led to the discovery that mutations in myosin VIIA are associated with *USH1B* and two forms of nonsyndromic deafness, *DFNB2* and *DFNA11* that map to the same region (Weil et al. 1995; Liu et al. 1997a,b).

A positional cloning approach also led to the discovery that a mutation in myosin VI (*Myo6*) causes deafness in the Snell's waltzer (*sv*) mouse (Fig. 3) (Avraham et al. 1995). Using fluorescent in situ hybridization (FISH), human myosin VI was mapped to chromosome 6q13 (Avraham et al. 1997). As this gene is an attractive candidate for human deafness, we refined the map position of human myosin VI (*MYO6*) by radiation hybrid mapping and characterized the genomic structure of myosin VI (Ahituv et al. 2001). Human myosin VI is composed of 32 coding exons, spanning a genomic region of approximately 70 kb (Fig. 4). Most recently, the *DFNA22* locus was mapped to this region. Subsequent mutation analysis showed that a missense mutation in the myosin VI gene is associated with nonsyndromic progressive hearing loss in an Italian family (Melchionda et al. 2001).

Shaker2 (*sh2*) mice, which exhibit recessive deafness, head-tossing and circling behavior, were created upon exposure to X-ray irradiation. Using an in vivo complementation approach by injecting bacterial artificial clones (BAC) that spanned the mapped mutated region into *sh2/sh2* fertilized eggs, the disease-causing gene, unconventional myosin XV, was discovered (Probst et al. 1998). The human deafness locus, *DFNB3,* lies in the homologous region, 17p11.2. Mutations in myosin XV were subsequently discovered in DFNB3 deaf individuals (Wang et al. 1998).

While we cannot determine the fate of the hair cells in humans with myosin VI-, VIIA-, and XV-associated deafness, we can make predictions by studying the mutants with unconventional myosin mutations. Examination of stereocilia of *sh1* mutants by SEM revealed severe disorganization of these structures, with small clusters arranged in abnormal orientations (Self et al. 1998). SEM and TEM on *sv* mice cochleas revealed severe stereociliar disorganization, with stereocilia fusing shortly after birth, suggesting a role for myosin VI in the anchoring of the apical hair cell membrane to the actin-rich cuticular plate (Fig. 5A and B) (Self et al. 1999). *sh2* stereocilia have an altogether different appearance; they are shorter than their normal counterparts. Taken together, the phenotype of the stereocilia in these three mutants demonstrates that the myosin VI, VIIA, and XV molecules have different and distinct roles in hair cell development and function (Fig. 5C). Though these roles may overlap, they are each clearly essential molecules, since each is required for normal cochlear function.

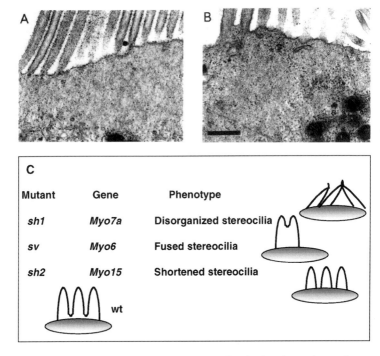

Fig. 5. A Transmission electron microscopy of cuticular plate at base of stereocilia in hair cell of 1-day-old wild-type mouse (Self et al. 1999). B Stereocilia
fusion in Snell's waltzer hair cells. *Bar*=500 nm. C Two other mouse mutants
with mutations in unconventional myosins led to stereociliar abnormalities, as
shown schematically

5 POU Transcription Factors in Deafness

Mutations in the POU family of transcription factors leads to both
X-linked and autosomal dominant forms of human hearing loss. The
POU-domain transcription factor family was originally identified due to
amino acid similarity between the mammalian PIT1/GHF1, OCT1,
OCT2 and *Caenorhabditis elegans* UNC86 proteins, leading to the
coining of the POU acronym (Semenza 1999). This family of developmental regulators contains a bipartite DNA-binding domain that includes a 60 amino acid homeodomain (POU_{HD}) and a 70–80 amino acid

POU-specific domain (POU_S). High-affinity binding to DNA requires the addition of this POU-specific domain.

The *DFN3* locus was mapped on chromosome Xq21 in 1988 by linkage analysis; 7 years later, mutations were identified in the *POU3F4* gene lying in this region (de Kok et al. 1995). Individuals with these mutations, which vary from single nucleotide deletions to position effects outside the coding region, suffer from both conductive and sensorineural hearing loss. Furthermore, stapes fixation, an enlarged internal auditory meatus and hypoplasia of the cochlea, is often seen in these individuals. The pathogenesis of deafness in human POU3F4-associated deafness has been studied extensively by gene-targeted mutagenesis (Minowa et al. 1999; Phippard et al. 1999). In the knockout mice, a malformation of the temporal bone was found due to the enlargement of the internal auditory meatus. Hypoplasia of the cochlea was also found and a reduction of coiling was found in 90% of the *Pou3f4* mutant mice. Histological analysis of the cochlea showed a smaller spiral limbus and thinner and less adherent fibrocytes of the spiral ligament. However, the mice show only a mild hearing loss and instead of stapes fixation, a malformation in the footplate of the stapes was found. A head bobbing phenotype, characteristic of vestibular dysfunction, was also observed due to a constriction in the bony labyrinth of the superior semicircular canals.

A mutation in the *POU4F3* gene is associated with hearing loss in a large Israeli Jewish family, Family H. Affected members of Family H suffer from progressive autosomal dominant sensorineural hearing loss (Fig. 6). The genetic basis of progressive hearing loss in this family was studied with the intention of mapping and eventually identifying the mutated gene associated with hearing loss. This was a project that could potentially take years, but due to advantages offered by the mouse as a research tool and technical advances in human genomics, the *DFNA15* gene was cloned in just 1 year (Vahava et al. 1998). The first individual reported to lose his hearing prematurely was born in 1843 in Libya. Eventually the family migrated to Israel, stopping in Tunisia and Egypt along the way. A genetic linkage study was performed on 12 affected and 11 unaffected individuals. Since only 13 dominant loci were known at the time this study was initiated, we used markers flanking these loci to determine if deafness in Family H mapped to a chromosomal region already known to harbor a gene for hearing loss. Genotypes of markers

A

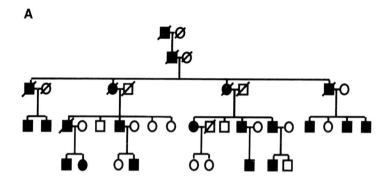

B

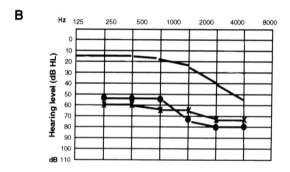

Fig. 6. A Family H pedigree showing autosomal dominant mode of inheritance
(Vahava et al. 1998). *Square*, male; *circle*, female; *black*, affected individual;
white, unaffected individual; *crossed*, deceased. **B** Auditory thresholds of an
affected member of Family H (Frydman et al. 2000)

on chromosome 5, near *DFNA1*, the first locus mapped for dominant,
nonsyndromic deafness (Lynch et al. 1997), suggested that deafness in
Family H was co-inherited with a gene near *DFNA1*, but not *DFNA1*
itself. Further genotyping confirmed linkage of deafness in Family H to
markers on chromosome 5q31-q33 proximal to *DFNA1* (Fig. 7). The
locus for deafness in Family H was named *DFNA15*, as it was the 15th
dominant locus identified. Examination of this chromosomal region in
humans did not reveal any interesting candidates. However, an ideal

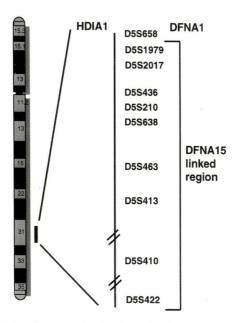

Fig. 7. The *DFNA15* locus was localized to a 25-cM region on human chromosome 5q31, proximal to the *DFNA1* locus

candidate was identified on mouse chromosome 18 in the region of homology to 5q31. There were two compelling reasons why the *Pou4f3* gene was a good candidate for deafness in Family H. First, targeted deletion of the entire *Pou4f3* gene leads to vestibular dysfunction and profound deafness in the knockout mice (Erkman et al. 1996; Xiang et al. 1997). Second, expression of murine *Pou4f3* is restricted to the cochlear and vestibular hair cells of the inner ear. Primers spanning the human gene were designed and DNA from both unaffected and affected Family H members was amplified. In deaf members of Family H, an 8-bp deletion (884del8) was identified in the 3′ portion of human *POU4F3* (Fig. 8A). This deletion leads to a frameshift, causing a stop codon to be formed prematurely (Fig. 8B). Presumably, a truncated protein is formed, which might still has the capacity to bind to DNA, though at a lower affinity due to the loss of a large portion of the POU homeodomain.

A

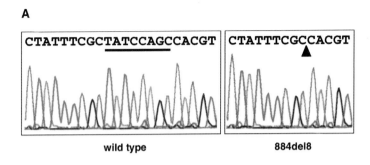

wild type   884del8

B **POU-specific domain**   **POU homeodomain**

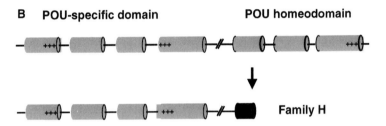

Family H

Fig. 8. A The POU4F3 mutation in Family H is shown by sequence analysis. Panel on the *left* shows sequence from unaffected individuals; panel on the *right* shows the 8-bp sequence (*underlined* in *left panel*) which is deleted in members of Family H with genetic hearing loss. **B** The schematic diagram shows the POU-specific domain (POU$_S$), and POU homeodomain (POU$_{HD}$) of the POU4F3 protein, and the truncated form of the POU4F3 protein formed by the 884del8 mutation in the POU$_{HD}$

We performed a clinical study of the auditory phenotype of affected members of Family H which indicated that *POU4F3* mutation-associated deafness cannot be identified through clinical evaluation, but only through molecular analysis (Frydman et al. 2000). Intrafamilial variability suggests that other genetic or environmental factors may modify the age of onset and rate of progression.

6 Gap Junction Proteins in Deafness

The most dramatic discovery in the field of the genetics of hearing loss has been the high incidence of mutations found in the gap junction protein, connexin 26 (locus designation, *GJB2*). Gap junction proteins encode the connexins, a component of connexons that allow molecules to pass from cell to cell. Connexin 26, which is expressed in the cochlea, is thought to have a role in the recycling of potassium ions back to the endolymph of the cochlear duct after the stimulation of the sensory hair cells (Kikuchi et al. 1995). Connexin 26, like other connexins, has four transmembrane domains, with the N- and C-termini in the cytoplasm. The molecular mass, which in the case of connexin 26 is 26 kDa, determines its nomenclature (Fig. 9B).

Three additional connexin genes have been implicated in deafness: connexin 30 (*GJB6*) and connexin 31 (*GJB3*) in NSHL, and connexin 32 (*GJB1*) in Charcot-Marie-Tooth disease (see connexins and deafness homepage, Table 1).

The *DFNB1* locus, which is located on chromosome 13q11–12, was the first deafness recessive locus to be discovered. *DFNB1* was initially identified by linkage analysis in a large Tunisian family exhibiting recessive hearing loss (Guilford et al. 1994). Genotyping of additional families eventually led to the identification of many more *DFNB1* families and the high prevalence of mutations at this locus became apparent once additional mutations in the connexin 26 gene were identified (Kelsell et al. 1997). Since then, connexin 26 (*GJB2*) mutations have been shown to account for about 30%–50% of autosomal recessive nonsyndromic deafness in regions all over the world (reviewed in Morell and Friedman 1999). Identification of mutations in this gene is facilitated by the fact that it is encoded by only one exon (Fig. 9A). One mutation, 35delG (also referred to as 30delG), accounts for the majority of mutant alleles. The 167delT mutation is the second most prevalent *GJB2* mutation, and so far, has only been reported in the Ashkenazi Jewish population. As many as 60 additional mutations have been identified in the connexin 26 gene.

We studied the prevalence and expression of inherited connexin 26 mutations associated with NSHL in the Israeli population (Sobe et al. 2000). The entire coding region of connexin 26 was sequenced in hearing-impaired children and adults in Israel in order to determine the

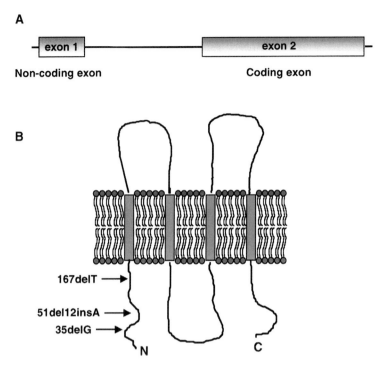

Fig. 9. A Connexin 26 is composed of two small exons, facilitating screening for mutations in this gene. **B** Schematic representation of the connexin 26 protein, with location of mutations found so far in the Israeli population

percentage of hearing loss attributed to connexin 26 and the types of mutations in this population. In a study including 75 individuals, 39% were found to have one or two of three connexin 26 mutations. The majority were either homozygote for 35delG and 167delT mutations or compound heterozygotes for both mutations (Fig. 9B). A novel mutation, involving both a deletion and insertion, 51del12insA, was identified in a family originating from Uzbekistan.

Examination of several parameters was made to establish whether genotype–phenotype correlations exist, including age of onset, severity of hearing loss and audiological characteristics (pure-tone audiometry, tympanometry, auditory brainstem response (ABR), and transient

35delG/35delG

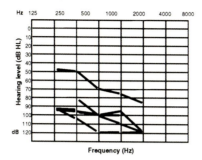

167delT/167delT

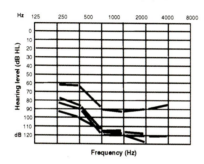

35delG/167delT

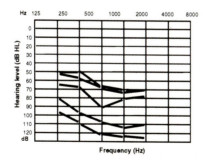

Fig. 10. Variability of hearing loss with different *GJB2* mutations (Sobe et al. 2000). Each threshold represents one individual (better ear is shown)

evoked otoacoustic emissions (TEOAE). All *GJB2* mutations were associated with prelingual hearing loss, though severity ranged from moderate to profound, with variability even among hearing-impaired siblings (Fig. 10). Our results are consistent with those in other populations in the world (Cohn et al. 1999; Denoyelle et al. 1999).

In a study of an American Ashkenazi Jewish population, a particularly high carrier rate (4.03%) for the 167delT mutation in the connexin 26 gene was reported (Morell et al. 1998). To determine the frequency in an Israeli Ashkenazi population, we screened 467 individuals for mutations in the connexin 26 gene (Sobe et al. 1999). In this population, we determined a carrier rate of 2.78% for the 167delT mutation. Although the carrier rate for the 167delT mutation within our sample group was somewhat lower than that reported previously, it was within the 95% confidence interval (2.5%–6.0%). This carrier rate is quite high when considering the situation in general of disease-carrying alleles in a population. Although deafness has not been cited as a common condition in Ashkenazi Jews, this carrier rate is comparable to, for example, Tay-Sachs disease (3%–4%), Gaucher disease (4%–6%), and Canavan disease (1%–2%) (Motulsky 1995), for which routine screening has been carried out for some years.

Unfortunately, the mouse knockout has not been informative with regards to human deafness. *GJB2* knockout homozygous mice show a lethal phenotype, dying at embryonic day 11 due to a dysfunction of the placenta, thought to be caused by decreased transplacental uptake of glucose (Gabriel et al. 1998). The difference in phenotype might be explained by the different morphology of the mouse and human placenta. The mouse placenta consists of two syncytiotrophoblast layers that are interconnected by gap junctions in order to support the transport of nutrients and removal of waste products, whereas the human placenta consists of only one large layer. A knockout directed only to the inner ear might address human connexin 26 pathology better.

7 The Identification of New Genes for Deafness

To date, 26 genes have been cloned that are known to be associated with human deafness. Despite the considerable progress over the past few years, it appears that there are many more genes to be discovered. A number of genes have been implicated in deafness in the mouse, and although they have not been associated with human deafness, the work has led to dramatic findings for auditory function. For example, the *Foxi1/Fkh10* knockout mice are deaf due to the fact that instead of an inner ear, the homozygous mutant mice contain an irregular and continuous cavity (Hulander et al. 1998). These mice also show a circling and head tilting behavior due to vestibular malfunction. The human *Fkh10* gene was mapped to chromosome 5q34, which implicates it as a possible deafness-causing gene if a nonsyndromic hearing loss family will be found linked to this region. The mouse jerker (*je*) circling and deafness phenotype is due to mutations in the espin gene, which encodes an actin-bundling proteins (Zheng et al. 2000). A role for this protein in inner ear stereocilia was previously unknown. The human espin gene maps to chromosome 1p36.11–36.31 and is another potential human deafness gene.

By injecting the alkylating agent ENU into mice, point mutations or small intragenic lesions can be created in a single gene to produce a large variety of phenotypes, including mouse mutants with abnormalities in sense organs, limbs, and the central nervous system (Hrabe de Angelis et al. 2000; Nolan et al. 2000) (for catalogue of ENU mouse mutants, see the GSF ENU-Mouse Mutagenesis Screen Project and the UK Mouse Genome Centre ENU Mutagenesis Programme; Table 1). In these screens, over 50 mouse mutants with deafness and vestibular dysfunction have been identified, providing a rich source of new mouse models that may be associated with human hearing loss.

8 Complexity in the Auditory System

Over 30 genes have been identified in human NSHL alone. More genes remain to be discovered, and their exact role in hearing elucidated. DNA microarray technology will enable scientists to understand the interaction of these many genes with one another. Thus far, proteins encoded by these genes include transcription factors, channel components, motor molecules, and extracellular matrix components. The inner ear is a complex organ, and thus it is no surprise that there is a requirement for a multitude of proteins to form its structures during development, to transduce sounds and to maintain proper balance. Ultimately, genetic research will facilitate detection of DNA variants that influence hearing loss, treatments to slow or reverse hearing loss (particularly in the elderly), and techniques to regenerate dying hair cells.

Acknowledgements. The author wishes to thank the many collaborators and laboratory members that have contributed to work described in this review. Research supported by the European Commission (QLG2–1999–00988), the NIH/Fogarty International Center Grant 1 R03 TW01108–01, the F.I.R.S.T. Foundation of the Israel Academy of Sciences and Humanities, and the Israel Ministry of Science, Culture & Sport.

References

Ahituv N, Sobe T, Robertson NG, Morton CC, Taggart RT, Avraham KB (2000) Genomic structure of the human unconventional myosin VI gene. Gene 261:269–275

Avraham KB, Hasson T, Steel KP, Kingsley DM, Russell LB, Mooseker MS. Copeland NG, Jenkins NA (1995) The mouse Snell's waltzer deafness gene encodes an unconventional myosin required for the structural integrity of inner ear hair cells. Nat Genet 11:369–375

Avraham KB, Hasson T, Sobe T, Balsara B, Testa JR, Skvorak AB, Morton CC, Copeland NG, Jenkins NA (1997) Characterization of unconventional MYO6, the human homologue of the gene responsible for deafness in Snell's waltzer mice. Hum Mol Genet 6:1225–1231

Cohn ES, Kelley PM (1999) Clinical phenotype and mutations in connexin 26 (DFNB1/GJB2), the most common cause of childhood hearing loss. Am J Med Genet 89:130–136

de Kok YJ, van der Maarel SM, Bitner-Glindzicz M, Huber I, Monaco AP, Malcolm S, Pembrey ME, Ropers HH, Cremers FP (1995) Association between X-linked mixed deafness and mutations in the POU domain gene POU3F4. Science 267:685–688

Denoyelle F, Marlin, S, Weil D, Moatti, L, Chauvin P, Garabedian E-N, Petit C (1999) Clinical features of the prevalent form of childhood deafness, DFNB1, due to a connexin-26 gene defect: implications for genetic counseling. Lancet 153:1298–1303

Erkman L, McEvilly RJ, Luo L, Ryan AK, Hooshmand F, O'Connell SM, Keithley EM, Rapaport DH, Ryan AF, Rosenfeld MG (1996) Role of transcription factors Brn-3.1 and Brn-3.2 in auditory and visual system development. Nature 381:603–606

Friedman TB, Sellers JR, Avraham KB (1999) Unconventional myosins and the genetics of hearing loss. Am J Med Genet (Sem Med Genet) 89:147–157

Frydman M, Vreugde S, Nageris BI, Weiss S, Vahava O, Avraham KB (2000) Clinical characterization of genetic hearing loss caused by a mutation in the *POU4F3* transcription factor. Arch Otolaryn Head Neck Surg 126:633–637

Gabriel H, Jung D, Butzler C, Temme A, Traub O, Winterhager E Willecke K (1998) Transplacental uptake of glucose is decreased in embryonic lethal connexin 26-deficient mice. J Cell Biol 140:1453–1461

Gibson F, Walsh J, Mburu P, Varela A, Brown KA, Antonio M, Beisel KW, Steel KP, Brown SDM (1995) A type VII myosin encoded by the mouse deafness gene shaker-1. Nature 374:62–64

Guilford P, Arab SB, Blanchard S, Levilliers J, Weissenbach J, Belkahia A, Petit C (1994) A non-syndromic form of neurosensory, recessive deafness maps to the pericentromeric region of chromosome 13q. Nat Genet 6:24–28

Hrabe de Angelis M, Flaswinkel H, Fuchs H, Rathkolb B, Soewarto D, Marschall S, Heffner S, Pargent W, Wuensch K, Jung M, Reis A, Richter T, Alessandrini F, Jakob T, Fuchs E, Kolb H, Kremmer E, Schaeble K, Rollinski B, Roscher A, Peters C, Meitinger T, Strom T, Steckler T, Holsboer F, Klopstock T, Gekeler F, Schindewolf C, Jung T, Avraham K, Behrendt H, Ring J, Zimmer A, Schughart K, Pfeffer K, Wolf E, Balling R (2000) Genome-wide, large-scale production of mutant mice by ENU mutagenesis. Nature Genet 25:444–447

Hulander M, Wurst W, Carlsson P, Enerback S (1998) The winged helix transcription factor Fkh10 is required for normal development of the inner ear. Nat Genet 20:374–376

Kelsell D, Dunlop J, Stevens HP, Lench NJ, Liang JN, Parry G, Mueller RF, Leigh IM (1997) Connexin 26 mutations in hereditary non-syndromic sensorineural deafness. Nature 387:80–83

Kikuchi T, Kimura RS, Paul DL, Adams JC (1995) Gap junctions in the rat cochlea: immunohistochemical and ultrastructural analysis. Anat Embryol (Berl) 191:101–118

Liu XZ, Walsh J, Mburu P, Kendrick-Jones J, Cope MJ, Steel KP, Brown SD (1997a) Mutations in the myosin VIIA gene cause non-syndromic recessive deafness. Nat Genet 16:188–190

Liu XZ, Walsh J, Tamagawa Y, Kitamura K, Nishizawa M, Steel K.P, Brown SDM (1997b) Autosomal dominant non-syndromic deafness caused by a mutation in the myosin VIIA gene. Nat Genet 17:268–269

Lynch E, Lee MK, Morrow JE, Welcsh PL, Leon PE, King MC (1997) Non-syndromic deafness DFNA1 associated with mutation of a human homolog of the *Drosophila* gene diaphanous. Science 278:1315–1318

Melchionda S, Ahituv N, Bisceglia L, Sobe T, Glaser F, Rabionet R, Lourdes Arbones M, Notarangelo A, Di Iorio E, Zelante L, Estivill X, Avraham KB, Gasparini P (2001) *MYO6*, the human homologue of the gene responsible for deafness in Snell's waltzer mice, is mutated in autosomal dominant nonsyndromic hearing loss. Am J Hum Genet 69:635–640

Mermall V, Post PL, Mooseker MS (1998) Unconventional myosins in cell movement, membrane traffic, and signal transduction. Science 279:527–532

Minowa O, Ikeda K, Sugitani Y, Oshima T, Nakai S, Katori Y, Suzuki M, Furukawa M, Kawase T, Zheng Y, Ogura M, Asada Y, Watanabe K, Yamanaka H, Gotoh S, Nishi-Takeshima M, Sugimoto T, Kikuchi T, Takasaka T, Noda T (1999) Altered cochlear fibrocytes in a mouse model of DFN3 nonsyndromic deafness. Science 285:1408–1411

Morell RJ, Friedman TB (1999) Deafness and mutations in the connexin 26 gene. N Engl J Med 340:1288

Morell RJ, Kim HJ, Hood LJ, Goforth L, Friderici K, Fisher R, Van Camp G, Berlin CI, Oddoux C, Ostrer H, Keats B, Friedman TB (1998) Mutations in the connexin 26 gene (*GJB2*) among Ashkenazi Jews with nonsyndromic recessive deafness. N Engl J Med 339:1500–1505

Motulsky A (1995) Jewish diseases and origins. Nat Genet 9:99–101

Nolan PM, Peters J, Strivens M, Rogers D, Hagan J, Spurr N, Gray IC, Vizor L, Brooker D, Whitehill E, Washbourne R, Hough T, Greenaway S, Hewitt M, Liu X, McCormack S, Pickford K, Selley R, Wells C, Tymowska-Lalanne Z, Roby P, Glenister P, Thornton C, Thaung C, Stevenson J-A, Arkell R, Mburu P, Hardisty R, Kiernan A, Erven A, Steel KP, Voegeling S, Guenet J-L, Nickols C, Sadri R, Naase M, Isaacs A, Davies K, Browne M, Fisher EMC, Martin J, Rastan S, Brown SDM, Hunter J (2000) A systematic, genome-wide, phenotype-driven mutagenesis programme for gene function studies in the mouse. Nature Genet 25:440–443

Phippard D, Lu L, Lee D, Saunders JC, Crenshaw BE III (1999) Targeted mutagenesis of the POU-domain gene Brn4/Pou3f4 causes developmental defects in the inner ear. J Neurosci 19:5980–5989

Probst FJ, Camper SA (1999) The role of mouse mutants in the identification of human hereditary hearing loss genes. Hear Res 130:1–6

Probst FJ, Fridell RA, Raphael Y, Saunders TL, Wang A, Liang Y, Morell RJ, Touchman JW, Lyons RH, Noben-Trauth K, Friedman TB, Camper SA (1998) Correction of deafness in shaker-2 mice by an unconventional myosin in a BAC transgene. Science 280:1444–1447

Self T, Mahony M, Fleming J, Walsh J, Brown SD, Steel KP (1998) Shaker-1 mutations reveal roles for myosin VIIA in both development and function of cochlear hair cells. Development 125:557–566

Self T, Sobe T, Copeland NG, Jenkins NA, Avraham KB, Steel KP (1999) Role of myosin VI in the differentiation of cochlear hair cells. Dev Biol 214:331–341

Semenza GL (1999) Transcription factors and human disease. Oxford Univ Press, Oxford

Sobe T, Erlich P, Berry A, Korostichevsky M, Vreugde S, Shohat M, Avraham KB, Bonné-Tamir B (1999) High frequency of the deafness-associated 167delT mutation in the connexin 26 (*GJB2*) gene in Israeli Ashkenazim. Am J Med Genet 86:499–500

Sobe T, Vreugde S, Shahin H, Davis N, Berlin M, Kanaan M, Yaron Y, Orr-Urtreger A, Frydman M, Shohat M, Avraham KB (2000) The prevalence and expression of inherited connexin 26 mutations associated with nonsyndromic hearing loss in the Israeli population. Hum Genet 106:50–57

Tymms MJ, Kola I (eds) (2000) Gene knockout protocols. Humana Press, New Jersey, USA

Vahava O, Morell R, Lynch ED, Weiss S, Kagan ME, Ahituv N, Morrow JE, Lee MK, Skvorak AB, Morton CC, Blumenfeld A, Frydman M, Friedman TB, King M-C, Avraham KB (1998) Mutation in transcription factor *POU4F3* associated with inherited progressive hearing loss in humans. Science 279:1950–1954

Wang A, Liang Y, Fridell RA, Probst FJ, Wilcox ER, Touchman JW, Morton CC, Morell RJ, Noben-Trauth K, Camper SA, Friedman TB (1998) Association of unconventional myosin MYO15 mutations with human nonsyndromic deafness DFNB3. Science 280:1447–1451

Weil D, Blanchard S, Kaplan J, Guilford P, Gibson F, Walsh J, Mburu P, Varela A, Levilliers J, Weston MD (1995) Defective myosin VIIA gene responsible for Usher syndrome type 1B. Nature 374:60–61

Xiang M, Gan L, Li D, Chen ZY, Zhou L, O'Malley BW Jr, Klein W, Nathans J (1997) Essential role of POU-domain factor Brn-3c in auditory and vestibular hair cell development. Proc Natl Acad Sci USA 93:11950–11955

Zheng L, Sekerkova G, Vranich K, Tilney LG, Mugnaini E, Bartles JR (2000) The deaf jerker mouse has a mutation in the gene encoding the espin actin-bundling proteins of hair cell stereocilia and lacks espins. Cell 102:377–385

6 Incontinentia Pigmenti:
The First Single Gene Disorder Due to
Disrupted NF-kappaB Function

S. Kenwrick

1 Incontinentia Pigmenti

1.1 Clinical Picture

Incontinentia pigmenti (IP) or Bloch-Sulzberger syndrome (Online Mendelian Inheritance in Man number 308300) is an X-linked dominant disorder characterised by male miscarriages and abnormalities of the skin, hair, teeth, nails eyes and nervous system in affected females (reviewed in Landy and Donnai 1993 and Table 1). It is estimated to affect between one in 10,000 and one in 100,000 women but as presentation of the disease is extremely variable, diagnosis is difficult in cases with no family history. Boys inheriting the defective gene usually die in utero before the end of the first trimester. In surviving girls, the disease presents in neonates and progresses through stages of inflammatory

Table 1. Clinical manifestations of incontinentia pigmenti

	Manifestation
Skin (100%)	Blistering
	Hyperkeratotic lesions
	Hyperpigmentation
	Dermal scarring
Teeth (80%)	Anodontia/hypodontia
	Malformed crowns
Hair (50%)	Alpecia
	Coarse hair
Nails (40%)	Dystrophy
	Subungual tumours
Eyes (40%)	Retinal dysplasia
Nervous system (10%–20%)	Epilepsy
	Spasticity
	Mental retardation
	Microcephaly

blistering, warty lesions and pigmented skin patches. The unsightly early skin lesions leave areas of under-vascularised scar tissue and patchy alopecia. Abnormalities of the teeth, nails and retina are also common. Retinal vessel dysplasia can lead to sight problems or retinal detachment and affected girls need regular ophthalmologic examinations. Some affected girls also have permanent neurological problems such as epilepsy, spasticity or mental retardation, which is a major concern for prospective mothers with the condition. Clinical presentation in IP families is extremely variable resulting in diagnostic uncertainty. Indeed, many cases are misdiagnosed as having neonatal infections and treated aggressively with inappropriate regimes. Identification of the responsible locus was therefore highly desirable in order that effective diagnosis and genetic counselling could be undertaken.

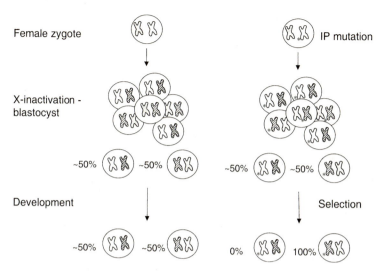

Fig. 1. X inactivation is skewed in incontinentia pigmenti (IP) female carriers. Lyonisation in the female blastocyst results in transcriptional inactivation of one X chromosome in each cell. In women carrying an IP mutation, subsequent selection against cells with the mutated X active leads to skewed X inactivation. The transcriptionally inactive Xs are *hatched* and the IP mutation is represented by an *asterisk*

1.2 Cells Expressing the Mutated X Chromosome Do Not Survive

The radical difference in presentation between male and female cases of IP is probably due to the presence of the gene on the X chromosome. Dosage compensation for X-linked genes is achieved by the process of transcriptional inactivation of one X in each cell (Lyonisation) in the developing female embryo. This process is random, but for 98% of women carrying an IP mutation there is subsequent selection against cells expressing the mutated X chromosome (Fig. 1). In tissues that have been examined so far i.e. blood and fibroblasts, selection occurs either in utero or around the time of birth so that, in adults, only cells expressing genes from the wild-type X chromosome are found (Parrish et al. 1996). Presumably this "clearing" of cells expressing the mutated gene explains the gradual resolution of the skin signs in IP female patients, diseased cells being gradually replaced by healthy ones. The resultant skewed pattern of X-inactivation is readily demonstrated by assays that detect methylation differences between active and inactive X chromosomes (Parrish et al. 1996; Woffendin et al. 1999).

1.3 The Genetics of IP

The clear X-linked dominant inheritance pattern of IP facilitated localisation of the gene. Over 10 years ago genetic linkage studies were used to establish that a major locus for IP resides in the Xq28 region of the long arm of the human X chromosome close to the gene for clotting factor VIII (Sefiani et al. 1989). Several laboratories then searched through large cohorts of IP families to identify meiotic recombination events that would define a small genomic interval containing the gene (Smahi et al. 1994; Parrish et al. 1996; Jouet et al. 1997). However, a paucity of crossovers between the IP gene and Xq28 markers meant that the interval was only narrowed to about 2 Mb of DNA at the telomeric end of the chromosome. This region contained many known genes but also a lot of uncharted DNA.

In 1996, to speed up identification of the relevant locus within this large region, five laboratories established a consortium whose sole purpose was to tease out the IP locus in Xq28. The consortium consisted of labs with complementary resources and expertise and their approach was to screen exons of candidate genes for mutations in a shared set of well-characterised, multigenerational families. By sharing the load, the consortium was able to eliminate many genes from its enquiries without redundancy. However, it wasn't until 1999 that the ideal gene was mapped to the region (Jin and Jeang 1999) and subsequently found to represent the IP locus (International Incontinentia Pigmenti Foundation 2000). This was the gene for NF-kappaB essential, modulator, or NEMO.

2 *NEMO* as a Candidate for the IP Locus

As its name suggests, NEMO is an essential component of the signalling pathway that involves the activation of Nuclear Factor-kappaB (NF-κB) (Rothwarf et al. 1998; Yamaoka et al. 1998). NF-κB is a family of dimeric transcription factors that are normally sequestered in an inactive state in the cellular cytoplasm by an inhibitory complex called IκB (containing subunits IκBα and β). In response to a variety of stimuli, including pro-inflammatory cytokines, this inhibitory complex is phosphorylated and targeted for degradation by proteosomes. This releases the NF-κB so that it can translocate to the nucleus and repress or activate genes with the right consensus sequence in their promoter (see Ghosh et al. 1998; Rothwarf and Karin 1999 for a review and Fig. 2). Phosphorylation of IκB is carried out by yet another kinase complex that contains at least two catalytic subunits (IKK1/α and IKK2/β) as well as NEMO (NEMO is also known as IKKγ, IKKAP and IKBKG).

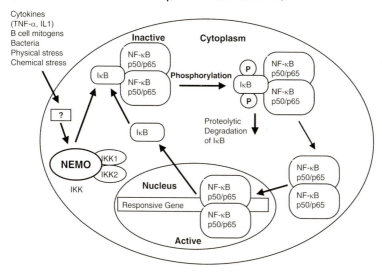

Fig. 2. The NF-κB signalling pathway. NF-κB transcription factor, represented here as a dimer comprising either p50 or p65 subunits, is sequestered in an inactive form in the cytoplasm by binding to an inhibitory complex (IκB). In response a variety of stimuli phosphorylation of this complex is undertaken by IKK, a complex containing catalytic subunits as well as regulatory subunit NEMO (IKKγ)

Although originally described as a signalling pathway in B lymphocytes, activation of NF-κB has now been found to be important for many
different cell types and cellular activities. NF-κB signalling is central to
many immune and inflammatory responses, such as those elicited by
pro-inflammatory cytokines (e.g. tumour necrosis factor alpha [TNFα]
and interleukin-1 [IL-1]). A component of this pathway was a very
attractive candidate for IP for the following reasons. Firstly, an early
inflammatory skin rash is seen in IP patients. Secondly, about 80% of IP
patients have abnormalities of tooth eruption or formation; NF-κB activation is important for osteoclast function and these cells are required to
eat away the alveolar bone before teeth can leave the gums. Thirdly,
NF-κB is an important regulator of apoptosis in many cell types and
abnormal regulation of apoptosis could account for the selective loss of
cells expressing the IP gene.

As is often the case in human genetics, an exceptional patient can
provide great insight and this has also been the case in the study of IP.
IP85m was the first reported affected son of an IP female to survive
more than 1 day. He not only had a skin rash and no teeth at the age of
2 years but also suffered from recurrent infections and showed signs of
osteopetrosis on bone biopsy (Mansour et al. 2001). These indicators of
immune deficiency, osteoclast malfunction and skin abnormality were
reminiscent of features observed in mouse lines with deficient NF-κB
activity (Attar et al. 1997; Franzoso et al. 1997; Hu et al. 1999; Li et al.
1999; Takeda et al. 1999).

3 *NEMO* Mutations

3.1 Most IP Cases Are Due to Genomic Rearrangement of *NEMO*

Screening the NEMO gene for mutations in IP cases was a daunting task
for two reasons. First, there was no genomic structure for the gene
available: the gene has been mapped to Xq28 on the basis of identification of 5′ untranslated sequence residing on a sequenced contig from
Xq28 (Jin and Jeang 1999). Second, because of X-inactivation skewing,
cDNA from IP cells would only contain wild-type *NEMO* transcripts.
The gene structure was resolved by sequencing large genomic (bacterial
artificial chromosome) clones and large polymerase chain reaction

(PCR) products and found to contain 12 exons of which 9 contained coding sequence (International Incontinentia Pigmenti Consortium 2000). This provided the data for screening individual exons of *NEMO*. The availability of some very rare cell lines expressing only the mutated X chromosome also enabled the consortium labs to amplify and sequence *NEMO* cDNA for a few cases. In two of these lines only one end (the 5′ end) of the *NEMO* transcript could be amplified by PCR indicating some aberration affecting the 3′ end of the gene. In a third line (IP85m) from the affected boy described above, a stop-codon mutation was found that would add 27 amino acids to the mature protein. This was found to occur de novo along with IP in this family and therefore very likely to be causing the condition. Interestingly, this mutation does not result in skewed X inactivation in the blood of the carrier mother and, unlike most IP mutations, is therefore compatible with cell survival (Mansour et al. 2001). The survival for 2.5 years of a boy carrying this mutation (IP85m) may, therefore, be a reflection of a "mild" effect on NEMO function. From these observations it appeared that the IP gene had been found. However, a systematic screen of *NEMO* exons identified only 4 additional intragenic mutations in 30 unrelated families. This anomaly was solved when it was discovered that in fact the majority of IP cases (80%) are due to a rearrangement of the gene. Recombination between two identical tandem copies of the MER67B repeat family, one inside intron 3 and one about 5 kb downstream of the last exon, results in deletion of exons 4–10 and a truncated mRNA species. Surprisingly, no exons appear to be missing when male cells carrying the deletion are examined. This is because a highly homologous pseudogene containing only exons 3–10 of NEMO is also present on the X chromosome (International Incontinentia Pigmenti Foundation 2000 and unpublished data).

IP therefore joins a growing list of disorders caused primarily by genomic recombination between highly homologous repeats (Lupski 1998) The origin of mutation for these examples shows a predilection for one parent. The common IP rearrangement occurs mostly during spermatogenesis rather than oogenesis (International Incontinentia Pigmenti Foundation 2000). As meiotic recombination does not occur between sex chromosomes outside of pseudoautosomal regions in males, the deletion primarily involves intra-chromosomal recombination between repeat sequences.

Screening for mutations in the 20% of patients without the common rearrangement is a current focus of the consortium, and to date many additional intragenic mutations have been identified, most of which would prematurely truncate the mature protein (IP consortium, unpublished data). This analysis is complicated by the presence of a homologous pseudogene and ways are being sought of preferentially identifying changes in the true gene. The variable nature of the clinical picture confounds genotype–phenotype comparisons for this disease. However, a cluster of mutations in exon 10 resulting from deletion or insertion at a cytosine tract do not always cause skewed X inactivation and sometimes allow male survival, in a manner analogous to the stop-codon mutation (Swaroop Aradhya, personal communication).

3.2 Diagnosis

The finding that 80% of IP mutations are identical has clear implications for diagnostic screening of this disease. A rapid PCR test has been developed to detect the boundary of genomic rearrangement that results from the common deletion, and any molecular diagnostic laboratory can apply this as a routine assay. This will enable rapid diagnosis and prenatal testing in 80% of cases of IP and prevent a great deal of inappropriate treatment.

3.3 Widening the Clinical Spectrum: IP and Hypohidrotic Ectodermal Dysplasia as Allelic Syndromes

The exceptional surviving IP male described above not only provided the first IP mutation but also allowed a vision of what phenotype "mild" mutations in *NEMO* may cause (see Sect. 2). This boy exhibited haematological disturbances, lymphoedema, capillary bed problems, sparse hair, skin problems (eczema, hyperpigmentation and absent sweat glands), immune deficiency, osteosclerosis and lack of tooth eruption This complex phenotype implicates the NF-κB pathway in a wide variety of cellular activities. It also has considerable overlap with a group of syndromes collectively known as hypohidrotic ectodermal dysplasias (HED). This genetically heterogeneous group of conditions involves abnormalities of tooth, hair and sweat gland development (Zonana 1993). Not surprisingly, *NEMO* mutations have now been described for males having HED with immunodeficiency (Zonana et al. 2000). Four mutations reported to date affect exon 10, i.e. reside close to the stop codon. In all four families these males are born of mothers that are heterozygous for the mutation but who themselves exhibit few if any signs of IP. Thus, *NEMO* mutations cause a very wide clinical spectrum ranging from male lethality with skewed X inactivation and variable signs in heterozygous females to male survival with immunodeficiency and random X inactivation in carrier females. For those mutations that allow male survival, clinical signs in carrier females may sometimes be insignificant.

4 NF-κB Function Is Compromised By *NEMO* Mutation

To examine the effect of *NEMO* mutations on NF-κB function, suitable patient cell lines are required. However, in most adult cases of IP, selection has eliminated cells expressing the mutated X and therefore these lines are not obtainable. Two cell lines obtained from spontaneously aborted foetuses have been found to only express a mutated X chromosome (International Incontinentia Pigmenti Foundation 2000 and Sect. 3.1). These lines carry the common deletion and have been used to study the effects of NEMO ablation. In cell extracts from both lines a truncated protein of about 21 kDa was found, probably representing the residual 133-residue product of the rearranged mRNA. NF-κB activation by pro-inflammatory cytokines and degradation of the inhibitory IκB complex was defective, indicating that this truncated product is not functional. Importantly, these fibroblast lines were more sensitive to tumour necrosis factor (TNF)α-induced cell death than control cells. This suggests a mechanism by which cells expressing this mutation in a heterozygous IP female are selectively lost.

5 NEMO-Deficient Mice Have an IP-Like Phenotype

Mouse lines without NEMO activity provide some insight into the cellular abnormalities resulting in IP. Three knockout lines have recently been described (Makris et al. 2000; Rudolph et al. 2000; Schmidt-Supprian et al. 2000). In all three cases, hemizygous males die in utero at embryonic day 12, which is analogous to the spontaneous abortions observed for the human disorder. In mice this is associated with a wave of hepatocyte apoptosis, a phenomenon that could also explain embryonic lethality in man. Female heterozygous mice also have a high mortality, unlike IP human carriers, but those that survive exhibit abnormalities of keratinocyte proliferation, apoptosis and differentiation. This results in areas of thickened epidermis. Pigment laden macrophage, hairlessness and infiltration of granulocytes into the skin are also reminiscent of the human phenotype. As with the human condition, the skin signs improve leaving only patchy areas affected. Ultimately, these models will provide a useful foundation for examining the contribution of NF-κB signalling to multiple tissues. The introduction into mice of mutations that do not completely ablate NEMO function, i.e. those that give rise to HED or mild IP phenotypes, may provide models for elucidating different aspects of NEMO and NF-κB function, such as a role in development of the lymphatic system.

6 Summary and Lessons

The effort to understand the underlying cause of incontinentia pigmenti has illustrated the advantage of combining the expertise of several laboratories in a scientific quest. The value of the Human Genome Project in providing candidates for disease is also highlighted by the story described here. The finding that the majority of cases of IP are caused by a genomic rearrangement and the complication of a neighbouring pseudogene reminds us that screening genes for mutation must not be restricted to the analysis of PCR products. The diagnostic benefit of a common mutation means that screening for potential cases of IP is simplified and will be introduced into many diagnostic centres.

The clinical phenotype observed for both IP females and surviving males will expand our knowledge of the influence of the NF-κB pathway on development and cellular function. This discovery will undoubt-edly generate wide scientific interest and further revelations concerning the link between NEMO and human disease over the next few years.

References

Attar RM, Caamano J, Carrasco D, Iotsova V, Ishikawa H, Ryseck RP, Weih F, Bravo R (1997) Genetic approaches to study rel/NF-κB/IκB function in mice. Semin Cancer Biol 8:93–101

Franzoso G, Carlson L, Xing LP, Poljak L, Shores EW, Brown KD, Leonardi A, Tran T, Boyce BF, Siebenlist U (1997) Requirement for NF-κB in osteo-clast and B-cell development. Genes Dev 11:3482–3496

Ghosh S, May MJ, Kopp EB (1998) NF-κB and rel proteins: evolutionary con-served mediators of immune responses. Annu Rev Immunol 16:225–260

Hu YL, Baud V, Delhase M, Zhang PL, Deerinck T, Ellisman M, Johnson R, Karin M (1999) Abnormal morphogenesis but intact IKK activation in mice lacking the IKKα subunit of IκB kinase. Science 284:316–320

International Incontinentia Pigmenti Foundation (France: Asmae Smahi GC, P Vabres, S Yamaoka, S Heuerz, A Munnich, A Israël; Germany: Nina Heiss, SM Klauck, P Kioschis, S Wiemann, A Poustka; Italy: Teresa Esposito, T Bardaro, F Gianfrancesco, A Ciccodicola, M D'urso; UK: Hayley Wof-fendin, T Jakins, D Donnai, H Stewart, S Kenwrick; USA: Swaroop Arad-hya, T Yamgata, M Levy, RA Lewis, D Nelson) (2000) Genomic rearrange-ment in NEMO impairs NF-κB activation and is a cause of incontinentia pigmenti. Nature 405:466–472

Jin DY, Jeang KT (1999) Isolation of full-length cDNA and chromosomal lo-calization of human NF- kappaB modulator NEMO to Xq28. J Biomed Sci 6:115–120

Jouet M, Stewart H, Landy S, Yates J, Yong SL, Harris A, Garret C, Hatchwell E, Read A, Donnai D, Kenwrick S (1997) Linkage analysis in 16 families with incontinentia pigmenti. Eur J Hum Genet 5:168–170

Landy SJ, Donnai D (1993) Incontinentia pigmenti (Bloch-Sulzberger syn-drome). J Med Genet 30:53–59

Li QT, Lu QX, Hwang JY, Buscher D, Lee KF, Izpisua-Belmonte JC, Verma IM (1999) IKK1-deficient mice exhibit abnormal development of skin and skeleton. Genes Dev 13:1322–1328

Lupski JR (1998) Genomic disorders: structural features of the genome can lead to DNA rearrangements and human disease traits. Trends Genet 14:417–422

Makris C, Godfrey V, Krähn-Senftleben G, Takahashi T, Roberts J, Schwartz T, Feng L, Johnson R, Karin M (2000) Female mice heterozygous for IKKγ/NEMO deficiencies develop a dermatopathy similar to the human X-linked disorder incontinentia pigmenti. Mol Cell 5:969–979

Mansour S, Woffendin H, Mitton S, Jeffery I, Jakins T, Kenwrick S, Murday V (2001) Incontinentia Pigmenti in a surviving male is accompanied by ectodermal dysplasia. Am J Med Genet 99:172–177

Parrish JE, Scheuerle AE, Lewis RA, Levy ML, Nelson DL (1996) Selection against mutant alleles in blood leukocytes is a consistent feature in Incontinentia Pigmenti type 2. Hum Mol Genet 5:1777–1783

Rothwarf D, Karin M (1999) The NF-κB activation pathway : a paradigm in information transfer from membrane to nucleus. http://www.stke.org/cgi/content/full/OC_sigtrans;1999/5/re1 26 Oct 1999

Rothwarf DM, Zandi E, Natoli G, Karin M (1998) IKK-γ is an essential regulatory subunit of the IκB kinase complex. Nature 395:297–300

Rudolph D, Yeh WC, Wakeham A, Rudolph B, Nallainathan D, Potter J, Elia AJ, Mak TW (2000) Severe liver degeneration and lack of NF-kappaB activation in NEMO/IKKgamma-deficient mice. Genes Dev 14:854–862

Schmidt-Supprian M, Bloch W, Courtois G, Addicks K, Israël A, Rajewsky K, Pasparakis M (2000) NEMO/IKKγ-deficient mice model Incontinentia Pigmenti. Mol Cell 5:981–992

Sefiani A, Abel L, Heuertz S, Sinnett D, Lavergne L, Labuda D, Hors-Cayla MC (1989) The gene for incontinentia pigmenti is assigned to Xq28. Genomics 4:427–429

Smahi A, Hyden-Granskog C, Peterlin B, Vabres P, Heuertz S, Fulchignoni-Lataud MC, Dahl N, Labrune P, Le Marec B, Piussan C et al (1994) The gene for the familial form of incontinentia pigmenti (IP2) maps to the distal part of Xq28. Hum Mol Genet 3:273–278

Takeda K, Takeuchi O, Tsujimura T, Itami S, Adachi O, Kawai T, Sanjo H, Yoshikawa K, Terada N, Akira S (1999) Limb and skin abnormalities in mice lacking IKKα. Science 284:313–316

Woffendin H, Jakins T, Jouet M, Stewart H, Landy S, Haan E, Harris A, Donnai D, Read A, Kenwrick S (1999) X-inactivation and marker studies in three families with incontinentia pigmenti: implications for counselling and gene localisation. Clin Genet 55:55–60

Yamaoka S, Courtois G, Bessia C, Whiteside ST, Weil R, Agou F, Kirk HE, Kay RJ, Israel A (1998) Complementation cloning of NEMO, a component of the IκB kinase complex essential for NF-kappaB activation. Cell 93:1231–1240

Zonana J (1993) Hypohidrotic (anhidrotic) ectodermal dysplasia: molecular genetic research and its clinical applications. Semin Dermatol 12:241–246

Zonana J, Elder ME, Schneider LC, Orlow SJ, Moss C, Golabi M, Shapira SK, Farndon PA, Wara DW, Emmal SA, Ferguson BM (2000) A novel X-linked disorder of immune deficiency and hypohidrotic ectodermal dysplasia is allelic to incontinentia pigmenti and due to mutations in IKK-gamma (NEMO). Am J Hum Genet 67:1555–1562

7 In Search of New Disease Models in the Mouse Using ENU Mutagenesis

C. Thaung, T. Hough, A.J. Hunter, R. Hardisty, P.M. Nolan

1 Introduction

Increases in the sophistication of screening procedures and in genome informatics have provided a unique opportunity to study mouse mutant models of human disorders. Advantages to using mouse models in the study of human disorders include the knowledge gained from comparative analysis of mouse and human genomes, the relatively short generation times needed to study inheritance in mice, the control of experimental variables in large populations of mice, unlimited access to adult and embryonic mouse tissues and the ease of manipulation of the mouse genome to study single gene mutations.

To date, three approaches have been used to identify and study mouse models of human disorders. Each approach is associated with its own merits and drawbacks, indicating that a combination of approaches might be the most rewarding. Since its introduction over 10 years ago, targeted mutagenesis has had a central role in the discovery of complex molecular pathways in mouse (Capecchi 1989) and has provided insight into many human biological processes. Although such studies were initially limited to analysis of genes with known sequence and function, recent developments have ensured that the phenotypic consequences of null mutations for any given gene sequence (Zambrowicz et al. 1998) as well as for subtle allelic variants of genes (Meyers et al. 1998) can readily be established. Quantitative trait locus (QTL) analysis in mice has been used extensively to study the genetic basis of complex human disorders such as atherosclerosis (Hyman et al. 1994) and epilepsy (Frankel et al. 1994). Although quantitative traits for many loci have been mapped, the identification and proof of identity of candidates has been difficult (Nadeau and Frankel 2000). The third approach, which will be described in detail here, is mutagenesis and screening for aberrant phenotypes in large populations of mice. Although this approach has been used for several decades, recent coupling of this approach with more specific and sophisticated screens has led to the identification of many novel mutations, several of which may have direct relevance in the study of human disorders. In this article, some of the developments in uncovering novel mouse mutations using this approach will be covered.

2 ENU Mutagenesis and Screening Programmes

2.1 ENU

N-ethyl-N-nitrosourea (ENU) has become the mutagen of choice for mutagenesis studies in mouse for several reasons. ENU has been shown to induce mutations in mice at a high frequency (0.0015 per locus per gamete). In effect, this means that, given the appropriate screening protocols, mutations at any given locus can be identified by screening fewer than 1,000 gametes (Hitotsumachi et al. 1985; Rinchik 1991). Unlike other mutagens, genetic lesions induced by ENU are primarily point mutations, meaning that observed phenotypes will be a consequence of single gene effects. One further advantage of using ENU is that it has the potential to induce all five types of allelic variants as they have been defined classically (amorphs, hypermorphs, hypomorphs, antimorphs and neomorphs). The generation of allelic series in this manner will not only provide a framework for the analysis of gene function but will also provide useful models for phenotypes that arise from subtle single nucleotide variants in humans.

ENU acts through ethylation of nucleic acids which results in mismatching of base pairs during DNA replication. The most common forms of mutations reported for the mouse are AT to TA transversions and AT to GC transitions and allelic variants include missense, nonsense and splice-site mutations (Noveroske et al. 2000). The mutants resulting from these variants include amorphs, hypomorphs, hypermorphs and antimorphs. The bias of ENU towards inducing mutations at TA base pairs would suggest that this approach might not allow one to recover all possible mutant alleles at a particular genetic locus. In these cases it may be necessary to use additional alkylating agents such as ethyl methanesulphonate that may be biased in favour of GC base pair mutations.

The mutagenic and cytotoxic effects of ENU vary depending on the dosage given and strain of animal used (Hitotsumachi et al. 1985; Davis et al. 1999). In addition, a particular mouse strain may be preferentially used for mutagenesis, as it may be susceptible or resistant to a particular trait being studied. The use of diverse protocols and strains should ensure the identification of independent classes of mutations. ENU induces a period of temporary sterility, during which time surviving mutagenised spermatogonial stem cells repopulate the testis. Generally, the effectiveness of mutagen treatment can be estimated on the basis of this temporary sterile period. We have found a fractionated dose of 2×100 mg/kg to be effective in inducing mutations in BALB/c male mice using these criteria and, moreover, find that a high rate of mutations are recovered using this dose (Nolan et al. 2000a). A comprehensive screen by Rinchik and Carpenter (1999) has indicated that duplication of mutations does not arise when an average of 29 G1 females (58 G1 mice) are analysed per G0 (mutagenised) male. In agreement with this, we have screened 50 G1 animals per mutagenised male and have not identified clonal duplications.

2.2 Screening

Having established efficient dose regimens for ENU, the design of multiple specific screens are required to identify as many mutant phenotypes as possible. Currently, many research centres have developed or are developing such screens (for a review of many of these screens see Vol. 11, Number 7 of Mammalian Genome). Reports of progress in two of these large-scale screens have been presented recently (Hrabe de Angelis et al. 2000; Nolan et al. 2000b). Screens range from simple dysmorphology screens to clinical chemistry and immunology screens. Screens undertaken at the Medical Research Council (MRC) Mammalian Genetics Unit, Harwell have incorporated research carried out at multiple research centres within the UK (MRC, Harwell; Queen Mary and Westfield College, London; Imperial College, London and SmithKline Beecham Pharmaceuticals, Harlow). This programme incorporates multiple phenotypic screens some of which will be described here.

The remit of the programme carried out by the UK consortium has been to undertake a large-scale screen for dominant ENU-induced mutations, confirm selected mutations by inheritance testing, and carry out low resolution mapping and functional analysis of particular mutants (Fig. 1). To date, over 1,100 phenotypes have been identified, 350 of these have been inheritance tested, over 160 of these are confirmed as inherited and, for more than 50, low-resolution map positions have been provided. Many of the mutants identified in our screen appear to represent mutations at novel loci, providing an important resource for functional genomics. Dominant genome-wide screens represent the simplest form of mutagenesis screen. With this approach, males are treated with ENU and subsequently mated with wild-type females (Fig. 1). Progeny from these crosses, which can each potentially carry mutations at up to 100 different loci, are then screened for abnormal phenotypes. Rather than screening for one particular phenotype, recent efforts have concentrated on conducting a wide range of phenotypic screens in the same population of F1 animals. This approach has dual benefits in that comprehensive phenotypic profiles can be generated for each potential mutant and that the maximum possible number of potential mutants per screening population can be identified.

Once abnormal phenotypes are identified in our phenotypic screens, they are retested by an independent observer after at least 1 week

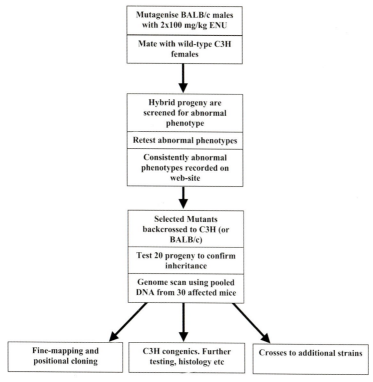

Fig. 1. Scheme for identification and inheritance testing of potential mutants

(Fig. 1). Only mice with consistently abnormal phenotypes are considered for inheritance testing. Generally, founder mice are backcrossed to wild-type C3H mice. If the phenotype is inherited in a dominant fashion, then 50% of backcross progeny should also express the abnormal phenotype. Often, however, we have found this number to be lower than 50% indicating that the abnormal phenotype may not be fully penetrant in the backcross being tested. If a mutation is confirmed as inherited, then additional backcross stock is generated and a low-resolution genome scan performed on DNA from 30 affected backcross mice. In cases where a phenotype is fully penetrant then DNA from unaffected backcross mice can also be used. Backcross mice can also be used for additional confirmatory tests (any tests that are not feasible for high-throughput screens) and for morphological analysis. Details of procedures used have been described elsewhere (Nolan et al. 2000a).

Several other screening programmes have been carried out or are being planned at this time. These include genome-wide recessive screens, region-specific recessive screens and sensitised screens. Genome-wide recessive screens require three generations of breeding and much animal house space. Moderate screens for embryonic lethal phenotypes (Kasarskis et al. 1998) as well as small-scale screens for recessive viable phenotypes (Hrabe de Angelis et al. 2000) have been fruitful. Region-specific screens represent a compromise between genome-wide dominant and recessive screens (Rinchik and Carpenter 1999). Using this approach, recessive mutations can be identified within a defined chromosomal region using a two-generation breeding scheme. This could be particularly useful in screening for mutations in regions where an important disease trait has been localised. Finally, sensitised screens can be used to identify or confirm genetic interactions between two molecules or to identify additional allelic variants at a particular locus. Such an approach has been used to identify novel alleles at the *kreisler* locus (Cordes and Barsch 1994).

3 Mutants Identified in Dysmorphology Screens

Dysmorphic mutants have been among the easiest to identify in mutagenesis screens. For the current dominant mutagenesis screen, mice are classified for dysmorphic anomalies from birth through to weaning and, in aged mice, at 6 and 12 months. Phenotypes that are routinely classified at each of these stages have been tabulated elsewhere (Nolan et al. 2000b) but these can briefly be grouped into several broad categories; growth, pigmentation, skin hair and vibrissae, tail, craniofacial, digits and limbs. To date, close to 30,000 mice have been screened for dysmorphic phenotypes and a summary of the results can be seen in Table 1. In total, over 1,100 phenotypes have been identified in the dominant screen. A complete list of individuals along with a brief description of each phenotype can be accessed via the MRC Harwell Web site http://www.mgu.har.mrc.ac.uk/mutabase/. Dysmorphic phenotypes represent close to half of the total number of phenotypes identified

Table 1. Summary of phenotypes and mutations identified in the screens for dominant mutations

	Number screened	Number of phenotypes	Observed phenotype rate	Number inheritance tested	Inherited mutations	Not inherited	Unknown
Growth	28,600	334	1 in 85	133	31	64	38
Pigment	28,600	140	1 in 204	41	21	14	6
Skin/Hair	28,600	70	1 in 409	21	10	5	6
Tail	28,600	43	1 in 665	9	4	3	2
Craniofacial	28,600	148	1 in 193	37	12	17	8
Digits/limbs	28,600	24	1 in 1192	14	3	8	3
Neurological/ Behavioural	15,000	526	1 in 28.5	139	56	52	31
Clinical Chemistry	1,900	58	1 in 33	28	9	7	12
Vestibular	28,600	54	1 in 530	20	15	3	2
Deafness	15,000	28	1 in 536	14	6	2	6
Eye/ Vision	6,000	81	1 in 74	41	24	10	7

in this screen and preliminary mapping data suggest that many of the mutants map to novel genetic loci. The observed phenotype rate for each category gives an indication of the average number of mice that require screening before an abnormal phenotype is identified. Several of the mice identified carry pleiotropic mutations with phenotypes in multiple dysmorphic categories (as well as anomalies in behavioural, sensory and clinical phenotypes). For example, mice with digit or limb anomalies often had associated craniofacial anomalies and several mice with white spotting also exhibited curled or looped tails. Approximately half of the abnormal phenotypes identified represent inherited single-gene mutations, although this value varies from category to category. Inheritance of a phenotype was recorded as unknown when testing data were not complete, the founder died or became ill or when inheritance of that particular phenotype had not been confirmed (Table 1).

3.1 Growth Anomalies

This category of mutations incorporates a broad spectrum of phenotypes including developmental growth retardation, nurturing defects, obesity, wasting disorders and bone disorders. Growth phenotypes are also dependent on environmental and social conditions (e.g. litter size, access to ad libitum diet, etc.). It is not surprising, therefore, that although 1 in 85 of the screening population has an abnormal growth phenotype, less than 25% of these represent inherited mutations (Table 1). Also, in many cases, the growth phenotype has been recorded as a secondary phenotype and its inheritance is marginal (hence the high number of unknowns). The growth mutant *sickly* (*Sic*) maps to mouse chromosome 7, apparently a novel locus. For this mutant, we have recorded a small phenotype which is evident from birth. Other features of this mutation have yet to be characterised. A second mutation, *nanomouse* (*Nano*) maps to the X chromosome and is lethal in hemizygous males. Females also exhibit a strikingly narrow face, prompting us to investigate whether there is an underlying bone disorder. Two additional growth mutants exhibit a phenotype which only becomes evident postnatally, and inheritance data suggest that these may be due to mutations in maternally imprinted genes. Late onset obesity and wasting phenotypes have also been recovered and these are currently in inheritance testing.

3.2 Pigmentation

This is an important class of mutation where the easily identifiable visible phenotype is often associated with pleiotropy. Coat colour mutants have led to important insights into erythropoiesis, deafness and vision, as well as into disorders such as obesity (Bedell et al. 1997; Jackson 1997). The majority of coat colour phenotypes represent inherited single-gene mutations (Table 1). Phenotypes include spotting, light and dark coat and skin pigmentation. Many of the mutants identified represent novel alleles at existing loci such as *Kit, Kitl, Mitf* and *Pax3*. However, from 41 phenotypes tested for inheritance, at least three of these map to novel loci. These include a dominant mutation with dark pigmented footpads, a belly spotting mutation mapping to a novel locus and a mutation with a variable phenotype including belly spot, curled tail and open neural tube at birth. The latter represents a mutation that could contribute towards the study of open neural tube phenotypes in humans and indicates that mutagenesis screens for single gene mutations has the potential to provide entry points into complex multigenic human disorders.

3.3 Skin/Hair/Vibrissae

Many of the phenotypes included in this group are X-linked phenotypes where X inactivation in heterozygous mutant females results in patterns of abnormal pigmentation or hair growth. Approximately 50% of phenotypes are inherited (Table 1). Inherited mutations include novel alleles at the *Nsdhl* (*Bpa*) and *Eda* (*Ta*) loci. Novel mutations include *saggy* (*Sagg*), a mutant line with loose skin that was identified when mice were scruffed. The mutation may represent a connective tissue anomaly.

3.4 Tail

This is another category of mutation where a readily identifiable phenotype may be indicative of an important biological process. Specifically, early embryonic development, neural tube closure and disorders of the axial skeleton have been associated with short curly and/or kinky tails. All three phenotypes have been detected in our screen for dominant phenotypes (Table 1). To date, only 9 phenotypes have been inheritance tested and approximately 50% of these are inherited. One of these has been confirmed as a novel allele at the *T* locus.

3.5 Craniofacial, Digit and Limb Defects

Genetic analysis of mouse mutants has provided many insights into craniofacial dysgenesis in humans (Bedell et al. 1997). Many of the digit/limb phenotypes identified in our screen also had craniofacial anomalies (Table 1). It is interesting to note that the rate of identification of craniofacial phenotypes is almost an order of magnitude higher than that for digit/limb defects. In addition, the low confirmed rate for inheritance of both classes of phenotype and the low penetrance of inherited mutations that we found underlies the genetic complexity of both of these processes. For this reason, many of the craniofacial phenotypes have been difficult to map genetically. By far the most frequent craniofacial phenotype identified is a shortened or bent face as in the *batface* and *yoda* mutants. Several other forms of craniofacial asymmetry are also seen, for example the *van gogh* (*Vng*) mutation has one ear lowered. This mutation maps to a novel locus on mouse chromosome 5 (Nolan et al. 2000b).

3.6 Neurological and Behavioural Mutants

Spontaneous neurological and behavioural mutations in mice have furthered our understanding of the physiology of both normal and diseased states (Hunter et al. 2000). Many of these spontaneous mutants had been identified through simple observational screens. Recently, the results of more complex behavioural screens in mice (Vitaterna et al. 1994) have proven that novel genes of important biological relevance can be uncovered. Bearing in mind that mouse mutants exist for only a small number of neurobiologically relevant genes, we decided to investigate whether the incorporation of systematic high-throughput screens could uncover novel neurological and behavioural mutations.

Although mice are continually screened for overt neurological and behavioural deficits from birth, we have found that systematic screening using quantitative and semi-quantitative measures can help in uncovering a far greater number of phenotypes. At 5 weeks, mice are assessed using the SHIRPA protocol (Rogers et al. 1997). Results from this screen provide a comprehensive profile and can identify anomalies in muscle and lower motorneuron, spinocerebellar, sensory, neuropsychiatric and autonomic function. The procedure is carried out in a simple testing arena and takes approximately 10 min per mouse. Two further high-throughput tests are added to assess behavioural function, assessment of locomotor activity (LMA) using cages equipped with beam-splitting monitors and prepulse inhibition (PPI) of the acoustic startle response (ASR) using custom-built startle chambers (Nolan et al. 2000a).

The use of semi-quantitative and quantitative phenotypic screens has produced a remarkable number of abnormal phenotypes (Table 1). Greater than 1 out of 30 animals subjected to these screens has a consistently abnormal phenotype. In addition, out of all phenotypes that were inheritance tested, greater than 50% represent inherited mutations. Many of the mutations exhibit incomplete penetrance and variable expressivity, indicating that most of the screened behaviours are multifactorial in nature. Inherited neurological and behavioural anomalies that have been detected include tremors, ataxia, fit-like behaviour, hyperactivity and aggression, abnormal gait (waddling and high-stepping), defective balance, poor muscle tone, low responsiveness to pain (toe pinch), high locomotor activity and exaggerated acoustic startle response. Several of these mutations map to novel loci including *robotic* (*Rob*), a neurological mutant with severe ataxia and cataracts, which maps to mouse chromosome 5 and *blind drunk* (*Bdr*), a mutant with abnormal scores in several SHIRPA parameters, which maps to mouse chromosome 2.

The neurological and behavioural screens have also led to the identification of novel alleles at existing loci. For example, two mutants with resting tremors mapped to the *Pmp22* locus on mouse chromosome 11 and point mutations in *Pmp22* coding sequence were identified in both of these mutants (Isaacs et al. 2000). PMP22 is associated with peripheral nerve myelination and PMP22 alterations are associated with many forms of inherited hereditary peripheral neuropathies including Charcot-Marie-Tooth type 1A disease, Dejerine Sottas syndrome and hereditary neuropathy with liability to pressure palsies. One of the novel alleles identified in our screen alters the same amino acid as has been found in patients with Dejerine Sottas syndrome and the other results in a truncation of the protein by seven amino acids. In addition, based on comparisons of several parameters in the SHIRPA semi-quantitative assessment, we have been able to establish that the former mutant is more severe than the latter, an observation also borne out by the electron microscopic analysis of the two mutations. This finding is encouraging, as it will enable us to compare the severities of allelic series of neurological and behavioural mutations. In doing so, SHIRPA analysis of allelic series may also aid in identifying functionally relevant domains in genes that have not been characterised to this extent in humans or mammals.

4 Mutants with Abnormal Clinical Chemistry Profiles

Advances in diagnostic techniques and the incorporation of automated screening methods that require as little as 2 μl of sample for biochemical testing, have made extensive biochemical testing of mouse blood samples feasible. Even a very basic standard profile of biochemical tests can supply the researcher with a wealth of information pertaining to the physiological state of an animal. Test profiles can be tailored for the specific needs of each study and can be a very useful tool for unravelling the quantitative and qualitative effects of gene expression. Such is the case in mouse mutagenesis studies where routine testing of mutagenised mouse blood samples can facilitate rapid identification of mice with biochemical abnormalities. This may lead to the discovery of novel genes or better understanding of previously discovered genes involved in the biochemical pathways of inherited metabolic disorders (Miklos and Rubin 1996; Meisler 1996).

4.1 Blood Collection

The method for blood removal from laboratory mammals was reviewed by the BVA/FRAME/RSPCA/UFAW joint working group on refinement in 1993. This group concluded that the removal of the tail tip under analgesia is the preferred method for blood sampling in the mouse. However, blood samples are also commonly obtained by making a superficial cut into the lateral tail vein. A Thermacage (Datesand, U.K.) can be of great benefit to speed up the circulation of mice prior to bleeding which facilitates sampling. Routine biochemical testing is usually performed on plasma samples collected in tubes coated with a heparin-based anticoagulant. Plasma specimens have been shown to be preferable to serum specimens for routine clinical analysis (Ladenson et al. 1974). For further details of methods, refer to Nolan et al. (2000b).

4.2 Identification of Abnormal Phenotypes and Mutants

A wide variety of diagnostic kits are commercially available for investigating biochemical parameters. We use an Olympus AU 400 clinical chemistry analyser (Olympus Diagnostic Systems, UK) to perform a standard set of 17 tests.

These tests incorporate the following basic profiles: liver profile (ALT, AST, total protein and albumin); kidney profile (urea, creatinine, sodium, potassium and chloride); bone profile (calcium, inorganic phosphorus and alkaline phosphatase); lipid profile (total cholesterol, high-density lipoprotein [HDL] cholesterol and triglycerides); glucose and bicarbonate. To date we have screened nearly 2,000 F1 offspring of mutagenised mice (Table 1). Regarding the rate of observation of phenotypes for this screen, 1 in 33 mice screened is high, and this is possibly a reflection of the multiple tests used. Nevertheless, it is clear that the high number of observed phenotypes validates the use of these screens to identify mouse mutants. For lines where inheritance testing has been completed, we have found that over half of the abnormal blood phenotypes are inherited.

Through the lipid profile studies, several mice with abnormal ratios of blood lipids have been identified. Confirmed inherited mutations include one line with high plasma triglycerides and two lines with low total and HDL cholesterol. Although there are differences in lipoprotein metabolism between mouse and man, many findings in mouse mimic those in humans. In mice and humans, HDL cholesterol levels are inversely proportional to the formation of atherosclerotic lesions (Paigen et al. 1994). HDL cholesterol is involved in reverse cholesterol transport from the peripheral tissues to the liver. Low levels of HDL cholesterol are, therefore, associated with an increased risk of atherosclerosis. Total cholesterol may also decrease with anaemia and malabsorption disorders. Hypertriglyceridaemia is a less important risk for coronary heart disease, and may result as a consequence of diabetes mellitus, hypothyroidism and liver disease (Carola et al. 1992; Marshall 1997).

To date a few mice with abnormal plasma calcium and inorganic phosphate have been found, but these phenotypes have not been inherited. Two of our current models display low levels of alkaline phosphatase (ALP). These mice will soon be subjected to histopathological investigation and their relevance to human hypophosphatasia investigated.

Non-fasting blood glucose test results can often be quite varied and are less reliable than fasting blood glucose measurements for diagnosing and monitoring diabetes. However, the inclusion of an ordinary non-fasting blood glucose test in our standard profile of tests has been valuable for the identification of individuals with insulin deficiencies. This test has identified several mice with consistently high blood glucose levels, and inheritance has been confirmed in some cases. Studies are currently underway using the intraperitoneal glucose tolerance test to investigate these phenotypes further.

5 The Identification of Deaf and Vestibular Mouse Mutants

5.1 Screening for Deafness and Balance Phenotypes

Mice with severe vestibular dysfunction are easily identified at weaning by their head weaving, circling and hyperactive behaviour. These mice are easily noticed, even by the untrained observer, and this behaviour may be an indication that there is also a deafness phenotype, although this is not always the case. Mice are also subjected to a small series of tests specifically designed to assess balance as part of the SHIRPA protocol carried out at weaning (Nolan et al. 2000b). These include the reaching response and the contact righting response. During the reaching response, mice are held by the tail and their behaviour monitored.

◄──►

Fig. 2. A, B Cleared inner ears from a control mouse (**A**) and a deaf, head-shaking mouse with a dominant, spontaneous mutation (**B**). Note the reduction in size of the posterior and lateral canals (*arrows*), the latter also being wider in diameter. The cochlear duct is also shorter in this mouse (*). *LSC*, lateral semicircular canal; *PSC*, posterior semicircular canal; *SSC*, superior semicircular canal; *C*, cochlea. **C, D** 3D reconstruction of the endolymphatic compartments (lumen) of a control inner ear (**C**), and that of a deaf and circling insertional mutant at E14.5 (**D**). This insertional mutant does not have semicircular canals, just an expanded cyst-like compartment (*Cy*) and a large endolymphatic duct. This approach allows development of normal and abnormal inner ears be viewed. *ED*, endolymphatic duct; *ES*, endolymphatic sac. **E, F** Scanning electron micrographs showing disorganisation of stereocilia in the inner ear of a *shaker1* mouse. A control mouse (**E**) and an allele of *shaker1*, *Myo7a6J* (**F**). This demonstrates the level of detail obtainable using scanning electron microscopy (SEM) to examine the surface of the sensory patches in the ear. Image taken from Self et al. (1998), with kind permission of The Company of Biologists Limited. (Copyright 1998)

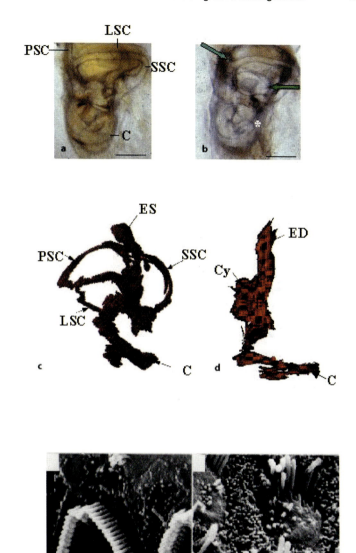

Fig. 2. Legend see p. 122

Wild-type mice will stretch out their front paws towards a horizontal surface, whereas mice with vestibular dysfunction will curl up towards their tail or rotate about their axis. During the contact righting response, mice are placed in a transparent tube, which is then rotated so the mouse is supine. Wild-type mice will rotate immediately to their original standing position, whereas mice with impaired vestibular dysfunction will walk upside down along this ventral surface, or in less extreme cases will show a delay in the righting reflex. These tests can be used to identify more subtle balance defects. To specifically test for deafness, mice are screened for a Preyer reflex. This is a flick of the pinna in response to a high frequency sound. A brief 20 kHz sound-burst at an intensity of 90 dB sound pressure level (SPL) is used (from a calibrated clickbox supplied by the MRC Institute of Hearing Research, Nottingham, UK). This is within the most acute hearing range for a mouse, which is around 18–24 kHz. The Preyer reflex is, however, a suprathreshold response, not an indication of normal thresholds, so it can best be used to identify mice with severe or profound hearing loss. Mice with mild or moderate hearing losses, without any other phenotypic manifestation (such as circling), may go undetected. The convenience and speed of using a clickbox counterbalances this possible "mis-scoring" of deafness phenotypes. More detailed assessment of the hearing ability of mice, such as round window recordings, would require specialist skills and equipment and are not suitable as a high-throughput screening tool. A small number of mice (10%), that have no obvious abnormal phenotype at weaning are kept and later tested (6 months and 1 year) as part of an aged cohort. This enables screening for late-onset deafness phenotypes.

Once mutants have been identified, the anatomy of the ear is investigated to ascertain the cause of the deafness or vestibular dysfunction.

The anatomy of the inner ear is not easy to investigate, not only because of its complexity but also because it is embedded in the dense temporal bone. However, it is relatively simple to dissect out whole inner ears and use clearing agents to visualise the bony structure of the ear. Otoconial membranes can also be visualised using this technique. Figure 2A shows a control, cleared inner ear, and Fig. 2B shows an inner ear from a mutant with a morphogenetic defect. This mouse has reduced lateral and posterior canals, with the lateral being wider than in controls. In order to examine the gross structure of the inner ear, paint filling of the labyrinth is employed (as described in Martin and Swanson 1993). To further examine the components of the inner ear, including the sensory patches and to study its development, 3-dimensional (3D) reconstruction of serial sections is a very useful, if time consuming, approach (See Fig. 2B and C). If the underlying pathology is more subtle and affects one or a few cell types within the sensory epithelium of the ear, then scanning electron microscopy (SEM) or transmission electron microscopy (TEM) are the favoured approaches (see Fig. 2E and F for an example of an SEM of stereocilia from organ of corti hair cells). In addition to the examination of the inner ear, the middle ear and its ossicles are also investigated. Ossicles can be dissected out from an adult mouse relatively easily and their structure examined.

5.2 Abnormal Phenotypes and Mutants

Despite the somewhat narrow phenotypic window available for deafness screening in mutagenesis programmes (due to the constraints of high throughput assessment of phenotype), the Harwell programme has identified 59 deaf and vestibular phenotypes, with a high percentage (~75%) being inherited mutations (Table 1). Some of the new ear mutants are listed in Table 2. The types of pathology identified so far have fallen into the following categories; middle ear defects, morphogenetic defects (where the gross structure of the inner ear is affected), defects of the sensory epithelium and late-onset deafness. This indicates that the screen has identified deaf and vestibular mutants with a range of pathologies and there is no bias towards any category in particular. In one case the mutation has already been identified in a molecule known to play a role in development of the neuroepithelium of the inner ear.

Deaf mouse mutants are very useful in the study of hereditary deafness in humans. The cochlea has an almost identical organisation in the two species and similar types of pathology are found in both mice and humans. Currently more than 50 loci have been identified for non-syndromic deafness in the human population. There is a deficiency in animal models for most of these conditions. This valuable mouse resource will help in closing this "phenotype gap" as well as contribute to our understanding of the pathways involved in auditory function in mice and humans.

Table 2. Deaf and vestibular mutants identified in the dominant screen

Mutant name (symbol)	Phenotype/site of pathology
Tornado (Tdo)	Circler
Orbitor (Obt)	Circler
Dizzy (Dz)	Circler
Cyclone (Cyn)	Circler
Ferris (Ferr)	Circler
Eddy (Edy)	Circler
Jeff (Jf)	Deaf only
Titania	Circler
Slalom (Slm)	Head shaker/patterning defect
Oberon	Deaf and circling
Metis	Circler
Leda	Circler
Spin cycle (Scy)	Circler
Pardon (Pdo)	Deaf only
Junbo	Late onset deafness
Tommy	Late onset deafness

6 Seeking Mouse Models of Inherited Eye Diseases

Inherited disorders are the leading cause of childhood blindness in the developed world. Genetic factors also play a part in common, later onset eye disease, including glaucoma and age-related macular degeneration, as well as systemic disorders with an ocular component, such as diabetes. Mouse models of these are useful in the search for the responsible human genes and allow dissection of the pathophysiology of these common diseases which result in considerable morbidity worldwide.

There are already many mouse models of inherited eye disease, some have occurred spontaneously (e.g. cataracts, retinal degeneration) and some have been induced, either randomly or as targeted mutations (e.g. Ehling 1985; Smith et al. 1997). One of the most frequently detected eye abnormalities has been cataract, presumably because it can be easily and rapidly screened for and because many genetic factors can contribute to correct lens protein structures, disruptions of which result in lens opacity. However, it is possible to study other eye abnormalities in the mouse, and models of human disease are becoming increasingly important for elucidating disease mechanisms and potential therapeutic targets (e.g. Frasson et al. 1999; Ali et al. 2000). Recently, an MRC-funded collaboration was set up between the MRC Human Genetics Unit, Edinburgh, and the MRC Mammalian Genetics Unit, Harwell. The aim of the collaboration is to identify mice from the mutagenesis programme with inherited abnormalities affecting any part of the eye or visual system, which could then lead to characterisation of the disease.

Clinical examination of a proportion of the mutagenised progeny was carried out using a slitlamp microscope (for anterior segment and lens) and a binocular indirect ophthalmoscope (for posterior segment). The dilating agent used was G. Tropicamide 1% (Chauvin Pharmaceuticals, UK) which has a full effect within 90 seconds. Due to the small size of the mouse eye and high curvature of the cornea and lens (resulting in a distorted view of the retinal periphery), examination was of necessity limited, but potentially detectable abnormalities would include corneal opacity, iris and pupillary irregularities, cataract and retinal degeneration.

6.1 Phenotypes Identified

Six thousand progeny have been examined, 41 of which showed eye abnormalities and were put into inheritance testing. A high proportion of these (58%) are inherited. The phenotypes are described in Table 3. Two groups of phenotype will be described further with reference to their applicability as models for known human diseases.

Table 3. Vision mutants identified in the dominant screen

Type of abnormality	Number placed into inheritance testing	Number showing inheritance
Corneal opacity	6	4
Pupillary irregularity	6	3
Cataract	8	4
Pigmentary retinal degeneration	8	7
Other retinal abnormality	10	3
Gross findings (multisystem)	3	3

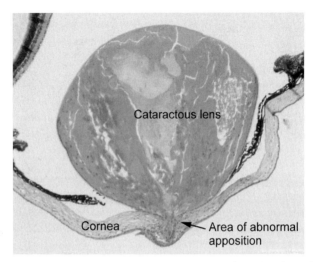

Fig. 3. Hematoxylin and eosin (H&E) stained section of eye showing adhesion of anterior lens to posterior cornea. The basement membranes are deficient centrally and the lens fibres remain attached to the corneal stroma. The corneal epithelium is heaped up over the abnormal layer

6.1.1 Example of an Anterior Segment Abnormality

A number of the eye mutants found in the programme have a very similar phenotype. Mutant eyes are small with corneal opacity, and there is often a central indentation of the cornea with a corresponding protrusion of the anterior lens. In one mutant line the pupils are dilated. Histology shows a central lens–corneal adhesion with local loss of the basement membranes, both of the anterior lens capsule and of Descemet's membrane in the posterior cornea (Fig. 3).

During normal embryonic development, the lens is derived from the lens vesicle, which forms an invagination of the surface ectoderm. Subsequently, the lens separates from the overlying surface ectoderm which forms the cornea. It is possible that in this series of mutants, the lens has failed to separate completely from the surface ectoderm. This phenotype is similar to that of *Pax6* (*Sey*) mutants. Three mutants from this screen have already been mapped to the *Pax6* region on mouse Chr2.

There are a number of human ocular disorders which are also related to an underlying PAX6 mutation, such as aniridia and Peters' anomaly (Prosser and van Heyningen 1998). Aniridia varies from a mild reduction in the iris stromal tissue to almost complete absence of the iris, while Peters' anomaly is characterised by a congenital defect in Descemet's membrane and the posterior corneal stroma, causing corneal opacity which can impair visual function and development. There may also be adhesions between the cornea and the iris and/or lens. These mouse mutants are phenotypically similar to the Peters' anomaly, and may provide information about the disease process and further insight into ocular development.

6.1.2 Example of a Pigmentary Retinopathy

Retinitis pigmentosa is a heterogeneous disease affecting approximately 1 in 3,000 people and is the most common cause of genetic blindness. Inheritance varies, with one study quoting 46% of cases as being isolated, 19% autosomal recessive, 19% autosomal dominant, 8% X linked and the remainder undetermined or mitochondrial (Bunker et al. 1984). Sufferers may have problems such as impaired night vision, visual field loss, and deteriorating vision associated with a characteristic pigmented appearance of the retina. The disease often progresses and may lead to legal blindness at an early age. Mutations in the human gene β-PDE, a cyclic guanosine monophosphate (GMP) phosphodiesterase involved in phototransduction, are responsible for the largest proportion of identified recessive retinitis pigmentosa mutations (McLaughlin et al. 1995).

One of the best-characterised mouse retinal degenerations is due to a mutation in the corresponding mouse gene encoding the β-subunit of cyclic GMP phosphodiesterase (*Pdeb*). The mutation was first described by Keeler in 1924 and it has since been retained in some laboratory inbred strains and wild populations. The mice screened in the Harwell programme are heterozygous for this mutation. The recessive mutation, $Pdeb^{rd1}$, carried by the C3H inbred strain, causes a fast retinal degeneration with 98% of the rod photoreceptor cells lost by postnatal day 17, although cone cells may remain functional until up to 18 months of age (Carter-Dawson et al. 1978). As the $Pdeb^{rd1}$ mouse has such a rapid onset of retinal cell death, it presents many problems as a model for investigating human disease mechanisms. However, mutagenesis may result in new mutations at the *Pdeb* locus resulting in less severe alleles.

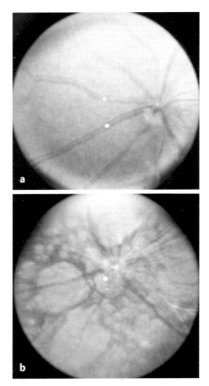

Fig. 4. A Photograph of a normal mouse retina. **B** The retina of a mouse with retinal degeneration, in this case *Pdeb^{rd1}*, although the other mutant lines develop an indistinguishable appearance over time. In contrast to the normal retina, the arterioles are narrowed and there are patches of depigmentation and hyperpigmentation throughout the fundus

Six founder mice have been identified from the Harwell mutagenesis programme which have retinal appearances similar to those of *Pdeb^{rd1}* homozygotes. Mating the founders to C3H (homozygous for the *Pdeb^{rd1}* mutation) produced litters where all of the offspring developed retinal degeneration, although the degeneration was delayed in some cases. Mapping studies currently underway indicate that these are six new recessive mutations of the *Pdeb* gene, and it is likely that these will include slower onset phenotypes than the original *Pdeb^{rd1}*. These new alleles of a well-characterised mouse gene, which is homologous to a human gene involved in disease, will give valuable information on mechanisms of retinal degeneration and cell death and possibly, in future, shed some light on therapy for this currently untreatable disease (Fig. 4).

7 Concluding Remarks

The generation and analysis of mouse mutations will be an important feature of post-genomics research. Recent evidence from large-scale phenotype-based screens has indicated that large numbers of single-gene mouse mutations can be identified using diverse screens of varying complexity. Screens can be devised with specific endpoints in mind, including those that might be useful in the study of human disorders. As the number of mutants in this valuable resource increases, the challenges will include the characterisation of molecular and genetic lesions associated with these mutations and investigation into their role in human physiological and diseased processes.

Acknowledgements. This work was supported by the Medical Research Council and SmithKline Beecham Pharmaceuticals. We are grateful for the contributions of scientists and staff at the MRC Mammalian Genetics Unit, Harwell, SmithKline Beecham Pharmaceuticals, Harlow, Imperial College London and Queen Mary and Westfield College, London. We thank R Arkell for helpful comments and discussions. K Arnold, S Cross, L McKie, K West and I Jackson contributed to the eye programme.

References

Ali RR, Sarra GM, Stephens C et al (2000) Restoration of photoreceptor ultrastructure and function in retinal degeneration slow mice by gene therapy. Nat Genet 25:306–310
Bedell MA, Largaespada DA, Jenkins NA, Copeland NG (1997) Mouse models of human disease, part II. Recent progress and future directions. Genes Dev 11:11–43
Bunker CH, Berson EL, Bromley WC et al (1984) Prevalence of retinitis pigmentosa in Maine. Am J Ophthalmol 97:357

Carola R, Harley JR, Noback CR (1992) Human anatomy and physiology, 2nd edn. McGraw-Hill, London

Carter-Dawson LD, LaVail MM, Sidman RL (1978) Differential effect of the rd mutation on rods and cones in the mouse retina. Invest Ophthalmol Vis Sci 17:489–498

Capecchi MR (1989) Altering the genome by homologous recombination. Science 244:1288–1292

Cordes SP, Barsh GS (1994) The Mouse segmentation gene *kr* encodes a novel basic domain-leucine zipper transcription factor. Cell 79:1025–1034

Davis AP, Woychik RP, Justice MJ (1999) Effective chemical mutagenesis in FVB/N mice requires low doses of ethylnitrosourea. Mamm Genome 10:308–310

Ehling UH (1985) Induction and manifestation of hereditary cataracts. Basic Life Sci 33:345–367

Frankel WN, Taylor BA, Noebels JL, Lutz CM (1994) Genetic epilepsy model derived from common inbred mouse strains. Genetics 138:481–489

Frasson M, Sahel JA, Fabre M et al (1999) Retinitis pigmentosa: rod photoreceptor rescue by a calcium-channel blocker in the rd mouse. Nat Med 5:1183–1187

Hitotsumachi S, Carpenter DA, Russell WL (1985) Dose repetition increases the mutagenic effectiveness of N-ethyl-N-nitrosurea in mouse spermatogonia. Proc Natl Acad Sci USA 82:6619–6621

Hrabe de Angelis M, Flaswinkel H, Fuchs H et al (2000) Genome-wide, large-scale production of mutant mice by ENU mutagenesis. Nat Genet 25:444–447

Hunter AJ, Nolan PM, Brown SDM (2000) Towards new models of disease and physiology in the neurosciences: the role of induced and naturally occurring mutations. Hum Molec Genet 9:893–900

Hyman RW, Frank S, Warden CH et al (1994) Quantitative trait locus analysis of susceptibility to diet-induced atherosclerosis in recombinant inbred mice. Biochem Genet 32:397–407

Isaacs AM, Davies KE, Hunter AJ et al (2000) Identification of two new Pmp22 mouse mutants using large-scale mutagenesis and a novel rapid mapping strategy. Hum Mol Genet 9:1865–1871

Jackson IJ (1997) Homologous pigmentation mutations in human, mouse and other model organisms. Hum Mol Genet 6:1613–1624

Kasarskis A, Manova K, Anderson KV (1998) A phenotype-based screen for embryonic lethal mutations in the mouse. Proc Natl Acad Sci USA 95:7485–7490

Keeler CH (1924) The inheritance of a retinal abnormality in white mice. Proc Natl Acad Sci USA 10:329–333

Ladenson JH, Tsai LB, Michael JM et al (1974) Serum versus heparinised plasma for eighteen common chemistry tests. Am J Clin Pathol 62:545–552

Marshall WJ (1997) Clinical chemistry, 3rd edn. Mosby, London

Martin P, Swanson GJ (1993) Descriptive and experimental analysis of the epithelial remodellings that control semicircular canal formation in the developing mouse inner ear. Dev Biol 159:549–558

McLaughlin ME, Ehrhart TL, Berson EL, Dryja TP (1995) Mutation spectrum of the gene encoding the beta subunit of rod phosphodiesterase among patients with autosomal recessive retinitis pigmentosa. Proc Natl Acad Sci USA 92:3249–3253

Meisler MH (1996) The role of the laboratory mouse in the human genome project. Am J Hum Genet 59:764–771

Meyers EN, Lewandoski M, Martin GR (1998) An *Fgf8* mutant allelic series generated by Cre- and Flp-mediated recombination. Nature Genet 18:136–141

Miklos GL, Rubin JM (1996) The role of the genome project in determining gene function: insights from model organisms. Cell 86:521–529

Nadeau JH, Frankel WN (2000) The roads from phenotypic variation to gene discovery: mutagenesis versus QTLs. Nat Genet 25:381–384

Nolan PM, Peters J, Vizor L et al (2000a) Implementation of a large-scale ENU mutagenesis program: towards increasing the mouse mutant resource. Mamm Genome 11:500–506

Nolan PM, Peters J, Strivens M et al (2000b) A systematic, genome-wide, phenotype-driven mutagenesis programme for gene function studies in the mouse. Nat Genet 25:440–443

Noveroske JK, Weber JS, Justice MJ (2000) The mutagenic action of *N*-ethyl-*N*-nitrosourea in the mouse. Mamm Genome 11:478–483

Paigen B, Plump AS, Rubin EM (1994) The mouse as a model for human cardiovascular disease and hyperlipidemia. Curr Opin Lipidol 5:258–264

Prosser J, van Heyningen V (1998) PAX6 mutations reviewed. Hum Mutat 11:93–108

Rinchik EM (1991) Chemical mutagenesis and fine-structure functional analysis of the mouse genome. Trends Genet 7:15–21

Rinchik EM, Carpenter DA (1999) *N*-ethyl-*N*-nitrosourea mutagenesis of a 6- to 11-cM subregion of the Fah-Hbb interval of mouse chromosome 7: completed testing of 4557 gametes and deletion mapping and complementation analysis of 31 mutations. Genetics 152:373–383

Rogers DC, Fisher EM, Brown SD et al (1997) Behavioral and functional analysis of mouse phenotype: SHIRPA, a proposed protocol for comprehensive phenotype assessment. Mamm Genome 8:711–713

Self T, Mahony M, Fleming J et al (1998) Shaker-1 mutations reveal roles for myosin VIIA in both development and function of cochlear hair cells. Development 125:557–566

Smith RS, Sundberg JP, Linder CC (1997) Mouse mutations as models for studying cataracts. Pathobiology 65:146–154

Vitaterna MH, King DP, Chang A-M et al (1994) Mutagenesis and Mapping of a Mouse Gene, *Clock*, Essential for Circadian Behavior. Science 264:719–725

Zambrowicz BP, Friedrich GA, Buxton EC, Lilleberg SL, Person C and Sands AT (1998) Disruption and sequence identification of 2,000 genes in mouse embryonic stem cells. Nature 392:608–611

Subject Index

Ernst Schering Research Foundation Workshop

Editors: Günter Stock
Monika Lessl

Vol. 16 (1995): Organ-Selective Actions of Steroid Hormones
Editors: D. T. Baird, G. Schütz, R. Krattenmacher

Vol. 17 (1996): Alzheimer's Disease
Editors: J.D. Turner, K. Beyreuther, F. Theuring

Vol. 18 (1997): The Endometrium as a Target for Contraception
Editors: H.M. Beier, M.J.K. Harper, K. Chwalisz

Vol. 19 (1997): EGF Receptor in Tumor Growth and Progression
Editors: R. B. Lichtner, R. N. Harkins

Vol. 20 (1997): Cellular Therapy
Editors: H. Wekerle, H. Graf, J.D. Turner

Vol. 21 (1997): Nitric Oxide, Cytochromes P 450,
and Sexual Steroid Hormones
Editors: J.R. Lancaster, J.F. Parkinson

Vol. 22 (1997): Impact of Molecular Biology
and New Technical Developments in Diagnostic Imaging
Editors: W. Semmler, M. Schwaiger

Vol. 23 (1998): Excitatory Amino Acids
Editors: P.H. Seeburg, I. Bresink, L. Turski

Vol. 24 (1998): Molecular Basis of Sex Hormone Receptor Function
Editors: H. Gronemeyer, U. Fuhrmann, K. Parczyk

Vol. 25 (1998): Novel Approaches to Treatment of Osteoporosis
Editors: R.G.G. Russell, T.M. Skerry, U. Kollenkirchen

Vol. 26 (1998): Recent Trends in Molecular Recognition
Editors: F. Diederich, H. Künzer

Vol. 27 (1998): Gene Therapy
Editors: R.E. Sobol, K.J. Scanlon, E. Nestaas, T. Strohmeyer

Vol. 28 (1999): Therapeutic Angiogenesis
Editors: J.A. Dormandy, W.P. Dole, G.M. Rubanyi

Vol. 29 (2000): Of Fish, Fly, Worm and Man
Editors: C. Nüsslein-Volhard, J. Krätzschmar

Vol. 30 (2000): Therapeutic Vaccination Therapy
Editors: P. Walden, W. Sterry, H. Hennekes

Vol. 31 (2000): Advances in Eicosanoid Research
Editors: C.N. Serhan, H.D. Perez

Vol. 32 (2000): The Role of Natural Products in Drug Discovery
Editors: J. Mulzer, R. Bohlmann

Vol. 33 (2001): Stem Cells from Cord Blood, In Utero Stem Cell Development, and Transplantation-Inclusive Gene Therapy
Editors: W. Holzgreve, M. Lessl

Vol. 34 (2001): Data Mining in Structural Biology
Editors: I. Schlichting, U. Egner

Vol. 35 (2001): Stem Cell Transplantation and Tissue Engineering
Editors: A. Haverich, H. Graf

Vol. 36 (2002): The Human Genome
Editors: A. Rosenthal, L. Vakalopoulou

Supplement 1 (1994): Molecular and Cellular Endocrinology of the Testis
Editors: G. Verhoeven, U.-F. Habenicht

Supplement 2 (1997): Signal Transduction in Testicular Cells
Editors: V. Hansson, F. O. Levy, K. Taskén

Supplement 3 (1998): Testicular Function:
From Gene Expression to Genetic Manipulation
Editors: M. Stefanini, C. Boitani, M. Galdieri, R. Geremia, F. Palombi

Supplement 4 (2000): Hormone Replacement Therapy
and Osteoporosis
Editors: J. Kato, H. Minaguchi, Y. Nishino

Supplement 5 (1999): Interferon:
The Dawn of Recombinant Protein Drugs
Editors: J. Lindenmann, W.D. Schleuning

Supplement 6 (2000): Testis, Epididymis and Technologies
in the Year 2000
Editors: B. Jégou, C. Pineau, J. Saez

Supplement 7 (2001): New Concepts in Pathology and Treatment
of Autoimmune Disorders
Editors: P. Pozzilli, C. Pozzilli, J.-F. Kapp

Supplement 8 (2001): New Pharmacological Approaches
to Reproductive Health and Healthy Ageing
Editors: W.-K. Raff, M. F. Fathalla, F. Saad